Structural Biology of DNA Damage and Repair

ACS SYMPOSIUM SERIES **1041**

Structural Biology of DNA Damage and Repair

Michael P. Stone, Editor
Vanderbilt University

Sponsored by the
ACS Division of Chemical Toxicology

American Chemical Society, Washington, DC

Library of Congress Cataloging-in-Publication Data

Structural biology of DNA damage and repair / Michael P. Stone, editor.
p. cm. -- (ACS symposium series ; 1041)
Includes bibliographical references and index.
ISBN 978-0-8412-2574-9 (alk. paper)
1. DNA repair--Congresses. 2. DNA damage--Congresses. I. Stone, Michael P.
QH467.S77 2010
572.8'6459--dc22
2010018421

The paper used in this publication meets the minimum requirements of American National Standard for Information Sciences—Permanence of Paper for Printed Library Materials, ANSI Z39.48n1984.

Distributed by Oxford University Press

PRINTED IN THE UNITED STATES OF AMERICA

Foreword

The ACS Symposium Series was first published in 1974 to provide a mechanism for publishing symposia quickly in book form. The purpose of the series is to publish timely, comprehensive books developed from the ACS sponsored symposia based on current scientific research. Occasionally, books are developed from symposia sponsored by other organizations when the topic is of keen interest to the chemistry audience.

Before agreeing to publish a book, the proposed table of contents is reviewed for appropriate and comprehensive coverage and for interest to the audience. Some papers may be excluded to better focus the book; others may be added to provide comprehensiveness. When appropriate, overview or introductory chapters are added. Drafts of chapters are peer-reviewed prior to final acceptance or rejection, and manuscripts are prepared in camera-ready format.

As a rule, only original research papers and original review papers are included in the volumes. Verbatim reproductions of previous published papers are not accepted.

ACS Books Department

Contents

Indexes

Preface

The human genome is continuously exposed to many classes of genotoxins. Of these, three that will be discussed herein include 5,6-dihydroxy-5,6-dihydrothymine (thymine glycol; Tg), O^6-methylguanine (O^6MeG), and benzo[*a*]pyrene. Thymine is highly susceptible to oxidative damage and Tg, forms by reaction of thymine with reactive oxygen species. Tg has been detected in animal and human urine; it is estimated that human cells repair hundreds of thymine glycol lesions per day (*1*, *2*). Studies on the structural biology of the Tg lesion (*3*) illustrate the role of configurational interconversion of DNA adducts plays upon modulating biological recognition and response. Purines are highly susceptible to alkylation and O^6MeG represents a frequent alkylation product in the genome (*4*). The O^6-alkyl transferases function to reverse the damage by transferring the alkyl group to a reactive active site cysteine residue (*5*). Polycyclic aromatic hydrocarbons such as benzo[*a*]pyrene are ubiquitous environmental pollutants; they are metabolized *in vivo* to genotoxic diol epoxides (B[*a*]PDE) (*6*). In all cases, if the genome is not repaired, these and other genotoxic lesions precipitate serious biological consequences, including altered gene expression, mutation, and cell death. In addition to the genotoxic responses, it is increasingly being recognized that DNA lesions can alter the epigenetic profiles that are imprinted by naturally occurring DNA modifications, e.g., cytosine methylation (*7*).

The development of site-specific approaches toward monitoring chemical lesions situated at a known position within DNA, e.g., with respect to their structures, and their replication (*8*, *9*), have led to tremendous advances in understanding of the chemical biology of DNA damage (*10*). Similar approaches are now facilitating site-specific studies of DNA repair, and provide insight as to how thermal destabilization of the damaged DNA facilitates lesion recognition. The structure of *E. coli* AlkA, which excises alkylated bases from DNA (*4*), complexed with DNA containing pyrrolidine, reveals a protein-induced DNA bending concomitant with flipping of the alkylated base (*11*, *12*). When bound to DNA containing 1-azaribose or ethenoadenine its human homologue AAG also discriminates between damaged and normal bases via base flipping (*13*, *14*). Base excision repair studies using DNA containing 8-oxoguanine complexed with FAPy glycosylases suggest how these oxidative damage lesions are recognized, and suggest the base-flipping mechanism by which excision occurs (*15–22*). Human uracil-DNA glycosylases also flip the damaged base out into the active site (*23*, *24*). Likewise, the structure of endonuclease IV complexed with an abasic site shows both the abasic site and its partner flipped out of the helix, thus promoting specific insertion and endonucleolytic cleavage of the abasic site within the active site pocket (*25*). Thus, the base flipping mechanism appears to be common to the base excision repair process (*26*, *27*). The first structural views of nucleotide excision repair also suggest the importance of base flipping into the active site pocket of the repair enzyme (*28*, *29*).

The impetus for this volume came from a recent symposium sponsored by the ACS Division of Chemical Toxicology, bringing together scientists interested in the synthesis and structures of site-specific DNA adducts and in mechanistic aspects of DNA repair. This volume presents work at the interface between the structural biology of DNA damage and DNA repair.

In closing, I gratefully acknowledge support from the American Chemical Society Division of Chemical Toxicology and the National Institute of Environmental Health Sciences R13 ES-016957.

References

1. Cathcart, R.; Schwiers, E.; Saul, R. L.; Ames, B. N. *Proc. Natl. Acad. Sci. U.S.A.* **1984**, *81*, 5633–5637.
2. Adelman, R.; Saul, R. L.; Ames, B. N. *Proc. Natl. Acad. Sci. U.S.A.* **1988**, *85*, 2706–2708.
3. Brown, K. L.; Adams, T.; Jasti, V. P.; Basu, A. K.; Stone, M. P. *J. Am. Chem. Soc.* **2008**, *129*, 11701–11710.
4. Lindahl, T.; Sedgwick, B.; Sekiguchi, M.; Nakabeppu, Y. *Annu. Rev. Biochem.* **1988**, *57*, 133–157.
5. Sedgwick, B. *Nat. Rev. Mol. Cell Biol.* **2004**, *5*, 148–157.
6. Szeliga, J.; Dipple, A. *Chem. Res. Toxicol.* **1998**, *11*, 1–11.
7. Baskunov, V. B.; Subach, F. V.; Kolbanovskiy, A.; Kolbanovskiy, M.; Eremin, S. A.; Johnson, F.; Bonala, R.; Geacintov, N. E.; Gromova, E. S. *Biochemistry* **2005**, *44*, 1054–1066.
8. Basu, A. K.; Essigmann, J. M. *Chem. Res. Toxicol.* **1988**, *1*, 1–18.
9. Moriya, M. *Proc. Natl. Acad. Sci. U.S.A.* **1993**, *90*, 1122–1126.
10. Delaney, J. C.; Essigmann, J. M. *Chem. Res. Toxicol.* **2008**, *21*, 232–252.
11. Hollis, T.; Ichikawa, Y.; Ellenberger, T. *EMBO J.* **2000**, *19*, 758–766.
12. Hollis, T.; Lau, A.; Ellenberger, T. *Prog. Nucleic Acid Res. Mol. Biol.* **2001**, *68*, 305–314.
13. Lau, A. Y.; Scharer, O. D.; Samson, L.; Verdine, G. L.; Ellenberger, T. *Cell* **1998**, *95*, 249–258.
14. Lau, A. Y.; Wyatt, M. D.; Glassner, B. J.; Samson, L. D.; Ellenberger, T. *Proc. Natl. Acad. Sci. U.S.A.* **2000**, *97*, 13573–13578.
15. Bruner, S. D.; Norman, D. P.; Verdine, G. L. *Nature* **2000**, *403*, 859–866.
16. Fromme, J. C.; Verdine, G. L. *Nat. Struct. Biol.* **2002**, *9*, 544–552.
17. Fromme, J. C.; Verdine, G. L. *J. Biol. Chem.* **2003**, *278*, 51543–51548.
18. Fromme, J. C.; Verdine, G. L. *EMBO J.* **2003**, *22*, 3461–3471.
19. Fromme, J. C.; Banerjee, A.; Huang, S. J.; Verdine, G. L. *Nature* **2004**, *427*, 652–656.
20. Banerjee, A.; Yang, W.; Karplus, M.; Verdine, G. L. *Nature* **2005**, *434*, 612–618.

21. Banerjee, A.; Santos, W. L.; Verdine, G. L. *Science* **2006**, *311*, 1153–1157.
22. Banerjee, A.; Verdine, G. L. *Proc. Natl. Acad. Sci. U.S.A.* **2006**, *103*, 15020–15025.
23. Parikh, S. S.; Mol, C. D.; Slupphaug, G.; Bharati, S.; Krokan, H. E.; Tainer, J. A. *EMBO J.* **1998**, *17*, 5214–5226.
24. Slupphaug, G.; Mol, C. D.; Kavli, B.; Arvai, A. S.; Krokan, H. E.; Tainer, J. A. *Nature* **1996**, *384*, 87–92.
25. Hosfield, D. J.; Guan, Y.; Haas, B. J.; Cunningham, R. P.; Tainer, J. A. *Cell* **1999**, *98*, 397–408.
26. Mol, C. D.; Hosfield, D. J.; Tainer, J. A. *Mutat. Res.* **2000**, *460*, 211–229.
27. Yu, B.; Edstrom, W. C.; Benach, J.; Hamuro, Y.; Weber, P. C.; Gibney, B. R.; Hunt, J. F. *Nature* **2006**, *439*, 879–884.
28. Min, J. H.; Pavletich, N. P. *Nature* **2007**, *449*, 570–575.
29. Scrima, A.; Konickova, R.; Czyzewski, B. K.; Kawasaki, Y.; Jeffrey, P. D.; Groisman, R.; Nakatani, Y.; Iwai, S.; Pavletich, N. P.; Thoma, N. H. *Cell* **2008**, *135*, 1213–1223.

Michael P. Stone

Department of Chemistry
Vanderbilt University
Nashville, TN 37235
(615) 322-2589 (telephone)
(615) 322-7591 (fax)
michael.p.stone@vanderbilt.edu (e-mail)

Chapter 1

Overview

Michael P. Stone*

Department of Chemistry, Vanderbilt University, Nashville, TN 37235
***michael.p.stone@vanderbilt.edu**

The chemical re-arrangement of initially formed DNA adducts may result in the formation of new types of damage having different genotoxic consequences than the originally formed adducts. Examples in this overview of the subject include adducts arising from aflatoxin B1, aldehydic electrophiles arising from cellular oxidative damage, and thymine glycol adducts arising from oxidation of thymine. In turn, such chemical transformations may modulate the recognition and repair of DNA damage by base excision repair and nucleotide excision repair processes, and may modulate the mutagenic processing of DNA damage.

Like the chameleon, DNA adducts sometimes change their coloration in the duplex. The chemical re-arrangement of initially formed adducts may result in the formation of new types of damage having different genotoxic consequences than the originally formed adducts.

Aflatoxin B_1 (AFB_1), a potent mycotoxin contaminating human foodstuffs, is amongst the most mutagenic environmental chemicals to which humans are exposed (*1*); it induces G→T mutations in *Escherichia coli* (*2*). The N7-deoxyguanosine adduct formed in DNA by AFB_1 epoxide (*3*, *4*) undergoes chemical re-arrangement to the corresponding formamidopyrimidine (FAPY) lesion. The significance of this chemistry with respect to the etiology of human cancer associated with this mycotoxin became apparent from two sets of observations. Early on, it was recognized that the FAPY form of the AFB_1 lesion exhibited a significantly longer half-life *in vivo* (*5*). The longer half-life would increase the potential for mutagenesis, but moreover, might suggest that the lesion is refractory to DNA repair. With respect to repair, AFB_1-FAPY lesions are preferentially repaired in *Escherichia coli* by nucleotide excision repair (NER) (*6*), whereas these lesions provide poor substrates for the MutM (*7*) FAPY glycosylase and its human analog 8-oxoguanine DNA glycosylase (*8*). Secondly, site-specific mutagenesis studies revealed that the FAPY re-arrangement product was likely to be the key genotoxic species responsible for the genesis of AFB_1-induced G→T mutations *in vivo* (*9*). Once formed, the AFB_1 FAPY adduct can undergo further re-arrangement between α and β anomers of the deoxyribose (*10*). Structural analyses of the α and β anomers suggest why the latter is thermodynamically more stable, and why the former may block DNA replication (*11*, *12*). While the β anomer induces the characteristic induced G→T mutations, the α anomer blocks replication in *Escherichia coli* (*9*, *10*).

Lipid peroxidation generates a number of aldehydic electrophiles (*13*). Among these, malondialdehyde (MDA) is a major product of lipid peroxidation; it is mutagenic in *Salmonella typhimurium* (*14*) and in mammalian cells (*15*). Both carbonyl equivalents in MDA are reactive, and it forms a pyrimidopurinone annelation product with N1 and N^2 of dG, abbreviated M_1dG (*16*). When placed complementary to cytosine in duplex DNA, M_1dG converts to the ring-opened derivative N^2-(3-oxo-1-propenyl)-dG. The ring-opening of M_1dG as a nucleoside or in oligodeoxynucleotides is a reversible second-order reaction with hydroxide ion (*17*). The reverse ring-closure mechanism involves rapid formation of protonated OPdG and 8-hydroxy-6,7-propenodG intermediates that slowly converts to M_1dG in a general-acid-catalyzed reaction (*18*). The significance of this chemistry with respect to mutagenesis was quickly recognized because the pyrimidopurinone lesion blocks Watson-Crick base pairing, whereas the ring-opened N^2-(3-oxo-1-propenyl)-dG aldehydic species allows Watson-Crick base pairing; thus, the position of this chemistry during DNA replication was anticipated to modulate the mutagenic response of the M_1dG lesion. In this regard, structures determined with the *Sulfolobus solfataricus* Y-family DNA polymerase Dpo4 revealed that M_1dG remained in the ring-closed state (*19*); significantly, this polymerase produced a spectrum of mutagenic products primarily involving incorporation of dATP opposite M_1dG during error-prone bypass.

The α,β-unsaturated aldehydes (enals) acrolein, crotonaldehyde, and 4-hydroxynonenal (4-HNE) are also endogenous byproducts of lipid peroxidation, arising as a consequence of oxidative stress (*20–23*). Addition of these enals to dG involves Michael addition of the N^2-amine to give N^2-(3-oxopropyl)-dG adducts, followed by cyclization of N1 with the aldehyde, yielding the corresponding 1,N^2-dG products. The principal acrolein adduct is γ-OH-PdG (*24*, *25*), although the regioisomeric 6-hydroxypyrimido[1,2-*a*]purin-10(*3H*)-one (α-OH-PdG) has also been observed (*25*, *26*). The γ-OH-PdG adduct exists as a mixture of C8-OH epimers. With crotonaldehyde, addition at N^2-dG creates a stereocenter at C6. Of four possible products, the two with the *trans* relative configurations at C6 and C8 are observed (*25*, *27*). These are also formed through the reaction of dG with two equivalents of acetaldehyde (*28–30*). The lipid peroxidation product 4-HNE afforded related dG-adducts (*31*). The corresponding 4-HNE-derived 1,N^2-dG adducts possess an additional stereocenter on the C6-sidechain, resulting in four observable diastereomers. The notion that acrolein, crotonaldehyde, and 4-HNE undergo similar chemistry was confirmed by the observation that γ-OH-PdG ring-opens to the *N*2-(3-oxopropyl)-dG aldehyde (**1**) when placed opposite dC (*32*). Site-specific mutagenesis reveals that these 1,*N*2-dG adducts induce predominantly G→T transversions in COS-7 cells (*33–35*). The significance of the ring-opening chemistry was recognized in the notion that the *N*2-(3-oxopropyl)-dG aldehydes could potentially undergo further chemistry in DNA to form interstrand cross-links. These interstrand cross-links occur in the 5′-CpG-3′ sequence context (*36–38*) and exist as equilibrium mixtures of carbinolamine, imine, or pyrimidopurinone species, favoring the carbinolamine linkage in duplex DNA (*38*, *39*). The aldehydes also yield peptide- and protein-DNA conjugates (*40*, *41*).

The thymine glycol adduct (Tg), 5,6-dihydroxy-5,6-dihydrothymine, is a readily formed oxidation product from thymine; it blocks DNA replication (*42*, *43*). It exists as two diastereomeric pairs of epimers, the 5*R cis*, *trans* pair (5*R*,6*S*;5*R*,6*R*) and the 5*S cis*, *trans* pair (5*S*,6*R*; 5*S*,6*S*). The 5*R* pair is more abundant. Structural studies indicate that the *cis*:*trans* ratio of epimers is modulated by the identity of the complementary base (*44*). In Chapter 1, Brown et al. examine the chemistry and structure of the 5*R*-Tg lesion in the 5′-GTgG-3′ sequence context, paired opposite either dA or dG (*44–46*). The 5*R*-Tg lesion is a substrate for base excision repair. This is mediated by at least two DNA *N*-glycosylase/AP lyases that are influenced by the diastereoisomer of Tg, the *cis–trans* epimerization of each diastereoisomer, and the identity of the complementary purine (*47*). The 5*R*-Tg lesion is also repaired by nucleotide excision repair (NER), although the effects of the *cis–trans* epimerization of each diastereoisomer of Tg with regard to NER have not been characterized. Both randomly introduced 5*R*-Tg and abasic sites are substrates for the *Escherichia coli* UvrABC proteins (*48*, *49*). Tg is also excised *in vitro* by human NER (*50*).

Alkylation of DNA bases produces a broad spectrum of cytotoxic and mutagenic lesions. Most alkylated bases are substrates for the base excision repair (BER) pathway (*51–53*). DNA glycosylases locate the damaged base and catalyze the hydrolysis of the C1′-N glycosyl bond, forming an abasic site, which is further processed by the sequential actions of an apurinic (AP) endonuclease, DNA polymerase and DNA ligase, repairing the DNA. DNA glycosylases have been characterized from eukaryotes (*54–58*), archaea (*59*), and bacteria (*60–63*). Alkylpurine DNA glycosylases recognize a diverse set of lesions, including 3-methyladenine (3mA), 7-methylguanine (7mG), and 1,*N*6-ethenoadenine (εdA). In Chapter 2, Rubinson et al. summarize the structures of alkylpurine DNA glycosylases determined to date. In the context of functional data these structures provide insight into the mechanisms of alkylpurine selection and excision.

One of the central questions in the field of DNA damage recognition concerns the basis by which a single damaged base can be recognized in a vast genome of undamaged, yet nearly isomorphic bases. Two distinct thermally driven intramolecular search mechanisms may be envisioned: one dimensional sliding and three dimensional hopping along the DNA chain (*64*, *65*). In Chapter 3, Friedman and Stivers summarize results indicating that the human BER enzyme uracil glycosylase (hUNG) exhibits dynamic fluctuations when bound to DNA that are consistent with transitioning between linear scanning and a pausing mode in which extrahelical thymine and uracil bases are interrogated. The absence of these motions in free UNG suggests that the enzyme uses the favorable free energy of DNA binding to loosen its own structure and activate dynamic modes necessary for the identification of damaged uracil bases.

Nucleotide excision repair (NER) represents the primary repair mechanism for the removal of bulky DNA adducts resulting from exposures to environmental genotoxins, including ultraviolet (UV) radiation (*66*, *67*). Defects in NER lead are associated with the disease *xeroderma pigmentosum* (XP), which is characterized by hypersensitivity to UV exposure and predisposition to developing skin cancers (*68*). The NER system recognizes and repairs many structurally disparate lesions and differs from BER in that the NER proteins are not lesion-specific (*69*, *70*). In global genome NER repair DNA lesions are identified by the XPC protein (*71*, *72*). Alternatively, in transcription-coupled NER, DNA lesions cause RNA polymerase to stall, eliciting the repair response (*73*). In global genome NER repair, it is believed that XPC senses helix destabilization rather than the lesion itself (*72*). In the NER process, following lesion identification, dual incisions are made at specific sites located on both sides of the damaged base, allowing excision of an oligodeoxynucleotide fragment containing the lesion. The resulting gap is then filled by a DNA polymerase. Although much is known regarding the biochemistry of XPC (*71*, *72*), less is known regarding of the underlying molecular mechanisms of its activity. Recently, however, a structure of the *Saccharomyces cerevisiae* XPC homolog, Rad4, was determined (*74*). In Chapter 4, Shell and Chazin review the structural biology of XPC and how the structure of Rad4 (*74*) influences the model of DNA damage recognition in global genomic NER. They compare and contrast the structure of Rad4 (*74*) with available biochemical data for the human XPC protein.

The methylation of DNA at CpG dinucleotides by methyltransferases (MTases) represents an epigenetic alteration of the genome that plays an important role in the regulation of gene expression in eukaryotes. Abnormalities in DNA methylation are associated with tumorigenesis. A specialized protein, O^6-alkylguanine-DNA alkyltransferase (AGT), recognizes and removes the alkyl group from O^6-alkyl-deoxyguanosine lesions in DNA. When AGT binds O^6-alkyl-deoxyguanosine (O^6-alkyl-dG) containing DNA, the protein interacts with the N1-, N^2-, and N7- positions of the guanine lesion (*75*). Following AGT binding to DNA, the adducted nucleotide is flipped into the active site of the protein, and the O^6-alkyl substituent is displaced via an S_N2-type mechanism, restoring normal guanine and producing alkylated protein. The role of DNA methylation in modulating AGT activity became apparent when it was observed that only 8% of lung tumors had G→A transitions in the *p53* gene when the promoter region of the gene coding for AGT was not methylated, thereby allowing protein expression (*76*). In contrast, 33% of tumors with a methylated AGT promoter had G→A mutations within the *p53* gene (*76*). Other human cancers that have low expression levels of AGT exhibit increased frequencies of G→A mutations in the *K-ras* proto-oncogene and the *p53* tumor suppressor gene as compared to tumors with normal expression of AGT (*77–79*). In Chapter 5, Guza et al. discuss the role of DNA sequence modulation with respect to the repair of O^6-alkyl-dG lesions by AGT. The rates of repair are influenced by DNA sequence context, secondary structure, and alkyl group identity. They conclude that differences in rates of repair between the different alkyl groups and different sequence contexts are not a result of difference in AGT binding and kinetics of nucleotide flipping, since these reaction steps are very fast and are unaffected by DNA sequence. In contrast, the rate of alkyl transfer is the slowest forward step in the repair of *O*6-Me-dG and appears to be dependent on the alkyl group identity and the local sequence context.

Polycyclic aromatic hydrocarbons (PAH) are carcinogens that are pervasive in the environment, resulting from the incomplete combustion of organic materials. Benzo[*a*]pyrene (B[*a*]P) is one of the most common (*80*). B[*a*]P is metabolized into 7,8-diol-9,10-epoxide stereoisomers, the major metabolite being the (+)-7*R*,8*S*,9*S*,10*R* enantiomer of 7*r*,8*t*-dihydroxy-*t*9,10-epoxy-7,8,9,10-tetrahydrobenzo[*a*]pyrene (B[*a*]PDE), although minor amounts of the (-)-7*S*,8*R*,9*R*,10*S* enantiomer also form (*81*). These B[*a*]PDE metabolites are reactive and predominantly alkylate the exocyclic amino group of guanine by *trans* addition to the C10 position of B[*a*]PDE, or by *cis* addition (*82*). The most extensively studied adducts are (+)-*trans* -B[*a*]P-*N*2-dG, (+)-*cis*-B[*a*]P-*N*2-dG, (-)-*trans*- B[*a*]P-*N*2-dG, and (-)-*cis*- B[*a*]P-*N*2-dG (*83*). These lesions are formed efficiently at guanines in 5′-CpG-3′ sequences that are recognition sites of mammalian MTases (*84*), and the efficiency of formation is enhanced in the presence of m5dC instead of dC (*84–86*). While the susceptibilities of different PAH diol epoxide-DNA adducts to removal by NER mechanisms have been examined (*87, 88*), and their mutagenic properties have been investigated (*84–86*), the potential impact of these lesions on DNA methylation status has received little attention. The fraction of methylated cytosines decreases in the DNA of mammalian cells treated with racemic B[*a*]PDE (*89, 90*). Utilizing biochemical approaches *in vitro*, Gromova et al. in Chapter 6 show that BPDE-derived DNA lesions alter DNA methylation by the prokaryotic MTases M.SssI and M.HhaI, and the mammalian MTase Dnmt3a, at 5′-CpG-3′ and 5′-GCGC-3′ sites. The data indicate that these effects depend on the conformational properties of the lesions and their positions within the DNA recognition sequence. The results suggest that B[*a*]PDE may initiate cancer not only by genotoxic mechanisms, but might also contribute to tumor development by epigenetic effects that involve changes in DNA methylation status.

References

1. Groopman, J. D.; Kensler, T. W.; Wild, C. P. *Annu. Rev. Pub. Health* **2008**, *29*, 187–203.
2. Foster, P. L.; Eisenstadt, E.; Miller, J. H. *Proc. Natl. Acad. Sci. U.S.A.* **1983**, *80*, 2695–2698.
3. Essigmann, J. M.; Croy, R. G.; Nadzan, A. M.; Busby, W. F., Jr.; Reinhold, V. N.; Buchi, G.; Wogan, G. N. *Proc. Natl. Acad. Sci. U.S.A.* **1977**, *74*, 1870–1874.
4. Smela, M. E.; Currier, S. S.; Bailey, E. A.; Essigmann, J. M. *Carcinogenesis* **2001**, *22*, 535–545.
5. Croy, R. G.; Wogan, G. N. *Cancer Res.* **1981**, *41*, 197–203.

6. Alekseyev, Y. O.; Hamm, M. L.; Essigmann, J. M. *Carcinogenesis* **2004**, *25*, 1045–1051.
7. Boiteux, S.; O'Connor, T. R.; Lederer, F.; Gouyette, A.; Laval, J. *J. Biol. Chem.* **1990**, *265*, 3916–3922.
8. Radicella, J. P.; Dherin, C.; Desmaze, C.; Fox, M. S.; Boiteux, S. *Proc. Natl. Acad. Sci. U.S.A.* **1997**, *94*, 8010–8015.
9. Smela, M. E.; Hamm, M. L.; Henderson, P. T.; Harris, C. M.; Harris, T. M.; Essigmann, J. M. *Proc. Natl. Acad. Sci. U.S.A.* **2002**, *99*, 6655–6660.
10. Brown, K. L.; Deng, J. Z.; Iyer, R. S.; Iyer, L. G.; Voehler, M. W.; Stone, M. P.; Harris, C. M.; Harris, T. M. *J. Am. Chem. Soc.* **2006**, *128*, 15188–15199.
11. Mao, H.; Deng, Z.; Wang, F.; Harris, T. M.; Stone, M. P. *Biochemistry* **1998**, *37*, 4374–4387.
12. Brown, K. L.; Voehler, M. W.; Magee, S. M.; Harris, C. M.; Harris, T. M.; Stone, M. P. *J. Am. Chem. Soc.* **2009**, *131*, 16096–16107.
13. Esterbauer, H. Aldehydic products of lipid peroxidation. In *Free Radicals, Lipid Peroxidation and Cancer*, McBrien, D. C. H., Slater, T. F., Eds.; Academic Press: London, 1982; p 101.
14. Mukai, F. H.; Goldstein, B. D. *Science* **1976**, *191*, 868–869.
15. Yau, T. M. *Mech. Ageing Dev.* **1979**, *11*, 137–144.
16. Basu, A. K.; O'Hara, S. M.; Valladier, P.; Stone, K.; Mols, O.; Marnett, L. J. *Chem. Res. Toxicol.* **1988**, *1*, 53–59.
17. Riggins, J. N.; Daniels, J. S.; Rouzer, C. A.; Marnett, L. J. *J. Am. Chem. Soc.* **2004**, *126*, 8237–8243.
18. Riggins, J. N.; Pratt, D. A.; Voehler, M.; Daniels, J. S.; Marnett, L. J. *J. Am. Chem. Soc.* **2004**, *126*, 10571–10581.
19. Eoff, R. L.; Stafford, J. B.; Szekely, J.; Rizzo, C. J.; Egli, M.; Guengerich, F. P.; Marnett, L. J. *Biochemistry* **2009**, *48*, 7079–7088.
20. Burcham, P. C. *Mutagenesis* **1998**, *13*, 287–305.
21. Chung, F. L.; Nath, R. G.; Nagao, M.; Nishikawa, A.; Zhou, G. D.; Randerath, K. *Mutat. Res.* **1999**, *424*, 71–81.
22. Chung, F. L.; Zhang, L.; Ocando, J. E.; Nath, R. G. *IARC Sci. Publ.* **1999**, *150*, 45–54.
23. Nair, U.; Bartsch, H.; Nair, J. *Free Radical Biol. Med.* **2007**, *43*, 1109–1120.
24. Galliani, G.; Pantarotto, C. *Tetrahedron Lett.* **1983**, *24*, 4491–4492.
25. Chung, F.; Young, R.; Hecht, S. S. *Cancer Res.* **1984**, *44*, 990–995.
26. Zhang, S.; Villalta, P.; Wang, M.; Hecht, S. *Chem. Res. Toxicol.* **2007**, *20*, 565–571.
27. Eder, E.; Hoffman, C. *Chem. Res. Toxicol.* **1992**, *5*, 802–808.
28. Hecht, S. S.; Upadhyaya, P.; Wang, M. *IARC Sci. Pub.* **1999**, *150*, 147–154.
29. Wang, M.; McIntee, E. J.; Cheng, G.; Shi, Y.; Villalta, P. W.; Hecht, S. S. *Chem. Res. Toxicol.* **2000**, *13*, 1149–1157.
30. Lao, Y.; Hecht, S. S. *Chem. Res. Toxicol.* **2005**, *18*, 711–721.
31. Winter, C.; Segall, H.; Haddon, W. *Cancer Res.* **1986**, *46*, 5682–5686.
32. de los Santos, C.; Zaliznyak, T.; Johnson, F. *J. Biol. Chem.* **2001**, *276*, 9077–9082.
33. Kanuri, M.; Minko, I.; Nechev, L.; Harris, T.; Harris, C.; Lloyd, R. *J. Biol. Chem.* **2002**, *277*, 18257–18265.

34. Fernandes, P.; Wang, H.; Rizzo, C.; Lloyd, R. *Environ. Mol. Mutagen.* **2003**, *42*, 68–74.
35. Fernandes, P.; Kanuri, M.; Nechev, L.; Harris, T.; Lloyd, R. *Environ. Mol. Mutagen.* **2005**, *45*, 455–459.
36. Kozekov, I. D.; Nechev, L. V.; Sanchez, A.; Harris, C. M.; Lloyd, R. S.; Harris, T. M. *Chem. Res. Toxicol.* **2001**, *14*, 1482–1485.
37. Kozekov, I. D.; Nechev, L. V.; Moseley, M. S.; Harris, C. M.; Rizzo, C. J.; Stone, M. P.; Harris, T. M. *J. Am. Chem. Soc.* **2003**, *125*, 50–61.
38. Stone, M. P.; Cho, Y.-J.; Huang, H.; Kim, H.-Y.; Kozekov, I. D.; Kozekova, A.; Wang, H.; Lloyd, R. S.; Harris, T. M.; Rizzo, C. J. *Acc. Chem. Res.* **2008**, *41*, 793–804.
39. Kim, H. Y.; Voehler, M.; Harris, T. M.; Stone, M. P. *J. Am. Chem. Soc.* **2002**, *124*, 9324–9325.
40. Kurtz, A.; Lloyd, R. *J. Biol. Chem.* **2003**, *278*, 5970–5976.
41. Sanchez, A.; Minko, I.; Kurtz, A.; Kanuri, M.; Moriya, M.; Lloyd, R. *Chem. Res. Toxicol.* **2003**, *16*, 1019–1028.
42. Ide, H.; Kow, Y. W.; Wallace, S. S. *Nucleic Acids Res.* **1985**, *13*, 8035–8052.
43. Clark, J. M.; Beardsley, G. P. *Nucleic Acids Res.* **1986**, *14*, 737–749.
44. Brown, K. L.; Adams, T.; Jasti, V. P.; Basu, A. K.; Stone, M. P. *J. Am. Chem. Soc.* **2008**, *129*, 11701–11710.
45. Brown, K. L.; Basu, A., K.; Stone, M. P. *Biochemistry* **2009**, *48*, 9722–9733.
46. Brown, K. L.; Roginskaya, M.; Zou, Y.; Altamirano, A.; Basu, A., K.; Stone, M. P. *Nucleic Acids Res.* **2009**, 428–440.
47. Ocampo-Hafalla, M. T.; Altamirano, A.; Basu, A. K.; Chan, M. K.; Ocampo, J. E.; Cummings, A., Jr.; Boorstein, R. J.; Cunningham, R. P.; Teebor, G. W. *DNA Repair (Amst)* **2006**, *5*, 444–454.
48. Lin, J. J.; Sancar, A. *Biochemistry* **1989**, *28*, 7979–7984.
49. Kow, Y. W.; Wallace, S. S.; Van Houten, B. *Mutat Res.* **1990**, *235*, 147–156.
50. Reardon, J. T.; Bessho, T.; Kung, H. C.; Bolton, P. H.; Sancar, A. *Proc. Natl. Acad. Sci. U.S.A.* **1997**, *94*, 9463–9468.
51. Fromme, J. C.; Banerjee, A.; Verdine, G. L. *Curr. Opin. Struct. Biol.* **2004**, *14*, 43–49.
52. Fromme, J. C.; Verdine, G. L. *Adv. Protein Chem.* **2004**, *69*, 1–41.
53. Huffman, J. L.; Sundheim, O.; Tainer, J. A. *Mutat. Res.* **2005**, *577*, 55–76.
54. Brent, T. P. *Biochemistry* **1979**, *18*, 911–916.
55. Karran, P.; Lindahl, T. *Biochemistry* **1980**, *19*, 6005–6011.
56. Chen, J.; Derfler, B.; Samson, L. *EMBO J.* **1990**, *9*, 4569–4575.
57. Berdal, K. G.; Bjoras, M.; Bjelland, S.; Seeberg, E. *EMBO J.* **1990**, *9*, 4563–4568.
58. Memisoglu, A.; Samson, L. *Gene* **1996**, *177*, 229–235.
59. Begley, T. J.; Haas, B. J.; Noel, J.; Shekhtman, A.; Williams, W. A.; Cunningham, R. P. *Curr. Biol.* **1999**, *9*, 653–656.
60. Riazuddin, S.; Lindahl, T. *Biochemistry* **1978**, *17*, 2110–2118.
61. Thomas, L.; Yang, C. H.; Goldthwait, D. A. *Biochemistry* **1982**, *21*, 1162–1169.
62. O'Rourke, E. J.; Chevalier, C.; Boiteux, S.; Labigne, A.; Ielpi, L.; Radicella, J. P. *J. Biol. Chem.* **2000**, *275*, 20077–20083.

63. Alseth, I.; Rognes, T.; Lindback, T.; Solberg, I.; Robertsen, K.; Kristiansen, K. I.; Mainieri, D.; Lillehagen, L.; Kolsto, A. B.; Bjoras, M. *Mol. Microbiol.* **2006**, *59*, 1602–1609.
64. Berg, O. G.; Winter, R. B.; von Hippel, P. H. *Biochemistry* **1981**, *20*, 6929–6948.
65. Halford, S. E.; Marko, J. F. *Nucleic Acids Res.* **2004**, *32*, 3040–3052.
66. Sancar, A. *Annu. Rev. Biochem.* **1996**, *65*, 43–81.
67. Sancar, A.; Lindsey-Boltz, L. A.; Unsal-Kacmaz, K.; Linn, S. *Annu. Rev. Biochem.* **2004**, *73*, 39–85.
68. Lehmann, A. R. *Biochimie* **2003**, *85*, 1101–1111.
69. Naegeli, H. *FASEB J.* **1995**, *9*, 1043–1050.
70. Wood, R. D. *Annu. Rev. Biochem.* **1996**, *65*, 135–167.
71. Sugasawa, K.; Ng, J. M.; Masutani, C.; Iwai, S.; van der Spek, P. J.; Eker, A. P.; Hanaoka, F.; Bootsma, D.; Hoeijmakers, J. H. *Mol. Cell* **1998**, *2*, 223–232.
72. Batty, D.; Rapic'-Otrin, V.; Levine, A. S.; Wood, R. D. *J. Mol. Biol.* **2000**, *300*, 275–290.
73. Fousteri, M.; Mullenders, L. H. *Cell Res.* **2008**, *18*, 73–84.
74. Min, J. H.; Pavletich, N. P. *Nature* **2007**, *449*, 570–575.
75. Spratt, T. E.; Wu, J. D.; Levy, D. E.; Kanugula, S.; Pegg, A. E. *Biochemistry* **1999**, *38*, 6801–6806.
76. Wolf, P.; Hu, Y. C.; Doffek, K.; Sidransky, D.; Ahrendt, S. A. *Cancer Res.* **2001**, *61*, 8113–8117.
77. Esteller, M.; Toyota, M.; Sanchez-Cespedes, M.; Capella, G.; Peinado, M. A.; Watkins, D. N.; Issa, J. P.; Sidransky, D.; Baylin, S. B.; Herman, J. G. *Cancer Res.* **2000**, *60*, 2368–2371.
78. Nakamura, M.; Watanabe, T.; Yonekawa, Y.; Kleihues, P.; Ohgaki, H. *Carcinogenesis* **2001**, *22*, 1715–1719.
79. Bello, M. J.; Alonso, M. E.; Aminoso, C.; Anselmo, N. P.; Arjona, D.; Gonzalez-Gomez, P.; Lopez-Marin, I.; de Campos, J. M.; Gutierrez, M.; Isla, A.; Kusak, M. E.; Lassaletta, L.; Sarasa, J. L.; Vaquero, J.; Casartelli, C.; Rey, J. A. *Mutat. Res.* **2004**, *554*, 23–32.
80. Luch, A. *Nat. Rev. Cancer* **2005**, *5*, 113–125.
81. Conney, A. H. *Cancer Res.* **1982**, *42*, 4875–4917.
82. Cheng, S. C.; Hilton, B. D.; Roman, J. M.; Dipple, A. *Chem. Res. Toxicol.* **1989**, *2*, 334–340.
83. Geacintov, N. E.; Cosman, M.; Hingerty, B. E.; Amin, S.; Broyde, S.; Patel, D. J. *Chem. Res. Toxicol.* **1997**, *10*, 111–146.
84. Denissenko, M. F.; Chen, J. X.; Tang, M. S.; Pfeifer, G. P. *Proc. Natl. Acad. Sci. U.S.A.* **1997**, *94*, 3893–3898.
85. Chen, J. X.; Zheng, Y.; West, M.; Tang, M. S. *Cancer Res.* **1998**, *58*, 2070–2075.
86. Weisenberger, D. J.; Romano, L. J. *J. Biol. Chem.* **1999**, *274*, 23948–23955.
87. Hess, M. T.; Gunz, D.; Luneva, N.; Geacintov, N. E.; Naegeli, H. *Mol. Cell Biol.* **1997**, *17*, 7069–7076.
88. Mocquet, V.; Kropachev, K.; Kolbanovskiy, M.; Kolbanovskiy, A.; Tapias, A.; Cai, Y.; Broyde, S.; Geacintov, N. E.; Egly, J. M. *EMBO J.* **2007**, *26*, 2923–2932.

89. Wilson, V. L.; Smith, R. A.; Longoria, J.; Liotta, M. A.; Harper, C. M.; Harris, C. C. *Proc. Natl. Acad. Sci. U.S.A.* **1987**, *84*, 3298–3301.
90. Wojciechowski, M. F.; Meehan, T. *J. Biol. Chem.* **1984**, *259*, 9711–9716.

Chapter 2

Structural Consequences of Epimerization of Thymine Glycol Lesions in Duplex DNA: Implications for DNA Repair

Kyle L. Brown,[1] Marina Roginskaya,[2] Yue Zou,[2] Alvin Altamirano,[3] Ashis K. Basu,[3] and Michael P. Stone[1,*]

[1]Department of Chemistry, Center in Molecular Toxicology, Vanderbilt-Ingram Cancer Center, Vanderbilt University, Nashville, TN 37235
[2]Department of Biochemistry and Molecular Biology, James H. Quillen College of Medicine, East Tennessee State University, Johnson City, TN 37614
[3]Department of Chemistry, University of Connecticut, Storrs, CT 06269
[*]michael.p.stone@vanderbilt.edu

Thymine glycol (Tg), 5,6-dihydroxy-5,6-dihydrothymine, forms in DNA by reaction of thymine with reactive oxygen species. It exists as two diastereomeric pairs of epimers, the 5*R* *cis*, *trans* pair (5*R*,6*S*;5*R*,6*R*) and the 5*S* *cis*, *trans* pair (5*S*,6*R*; 5*S*,6*S*). The 5*R* pair is more abundant. At 30 °C, a 70%:30% *cis*:*trans* ratio of epimers is present in this sequence when 5*R*-Tg is opposite dA. For the *cis* epimer Tg and A remain in the Watson-Crick alignment. The Tg N3H imine and A N^6 amine protons undergo increased solvent exchange. Stacking between Tg and the 3′-neighbor G•C base pair is disrupted. The solvent accessible surface and T_2 relaxation of Tg increases. Molecular dynamics calculations predict that the axial conformation of the Tg CH_3 group is favored; propeller twisting of the Tg•A pair and hydrogen bonding between Tg OH6 and the N7 atom of the 3′-neighbor guanine alleviate steric clash with the 5′-neighbor base pair. Tg also destabilizes the 5′-neighbor G•C base pair. Under these conditions, the human NER protein XPA binds to the 5*R*-Tg lesion comparably to the C8-dG acetylaminoflourene (AAF) adduct, whereas XPC/HR23B binding of the Tg lesion is superior than to the AAF adduct. In comparison, this lesion is processed by the *Escherichia coli* UvrA and UvrABC proteins less efficiently than the C8-dG AAF adduct. The destabilization of two base pairs by the *cis* epimer may facilitate flipping both base pairs from the helix, enabling XPC/HR23B binding prior to recruitment of XPA. When 5*R*-Tg pairs opposite dG in this sequence context only the *cis* epimer is observed. Tg assumes the wobble orientation and stacks below the 5′-neighbor dG, while the mismatched dG stacks below the 5′-neighbor dC. Stacking between Tg and the 3′-neighbor G•C base pair is disrupted. Differences in base excision repair of the Tg•G and Tg•A pairs by hNEIL1 may be related to the wobble orientation of the *cis* Tg epimer in the Tg•G pair, and the lack of hydrogen bonding between the Tg OH groups and the N7 atom of the 3′-neighbor dG. Hydrogen bonding between Tg^6 OH6→G^7 N7 in the Tg•A pair may increase the energetic barrier with regard to flipping of the Tg lesion into the active site pocket of the glycosylase, hindering repair.

Introduction

In DNA, 5,6-dihydroxy-5,6-dihydrothymine, known as thymine glycol (Tg), is formed by exposure to ionizing radiation, as well as a variety of chemical oxidizing agents (*1*, *2*). Tg is also formed by oxidation of 5-methylcytosine followed by hydrolytic deamination of 5-methylcytosine glycol (*3*, *4*). The C5 and C6 atoms in Tg are chiral and Tg exists in DNA as two diastereomeric pairs of epimers, the 5*R* *cis*, *trans* pair (5*R*,6*S*;5*R*,6*R*) and the 5*S* *cis*, *trans* pair (5*S*,6*R*; 5*S*,6*S*) (Scheme 1) (*5–7*). The 5*R* pair is more abundant and more stable and for both the 5*R* and 5*S* pairs, the *cis* epimers predominate at the nucleoside level (*6*). Tg has been detected in animal and human urine; human cells may repair hundreds of thymine glycol lesions per day (*8*, *9*).

If not repaired, the 5*R*-Tg lesion hinders DNA replication (*10*, *11*) and is lethal to cells (*12–16*). However, DNA polymerases lacking 3′→5′ exonuclease activity do bypass the lesion with limited efficiencies (*17–19*). The bypass of the lesion by Y-family DNA polymerases is stereospecific, with pol η bypassing the 5*R*-Tg lesion most efficiently (*20*) and pol κ bypassing the 5*S*-Tg lesion most efficiently (*21*).

Normally in duplex DNA thymine is paired opposite adenine. Consequently, oxidative damage to thymine results in the formation of Tg opposite dA in DNA. However, when 5-methylcytosine is oxidatively damaged, forming 5-methylcytosine glycol, hydrolytic deamination yields Tg mismatched with dG (*3*). Since >70% of cytosines in the 5′-CpG-3′ sequence context are believed to be methylated in mammalian cells (*3*), the formation of Tg arising from oxidative damage to 5-methylcytosine is anticipated to be biologically significant. If not recognized and repaired by the cell, the Tg•G mismatch is anticipated to result in a $C^{5\text{-}Me}$→T mutation.

A

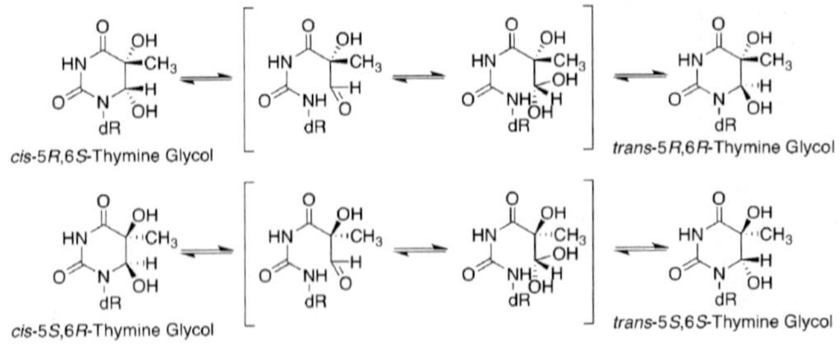

B

$5'$-d($G^{1}\ T^{2}\ G^{3}\ C^{4}\ G^{5}\ X^{6}\ G^{7}\ T^{8}\ T^{9}\ T^{10}G^{11}T^{12}$)-$3'$

$3'$-d($C^{24}A^{23}C^{22}G^{21}C^{20}A^{19}C^{18}A^{17}A^{16}A^{15}C^{14}A^{13}$)-$5'$

C

$5'$-d($G^{1}\ T^{2}\ G^{3}\ C^{4}\ G^{5}\ X^{6}\ G^{7}\ T^{8}\ T^{9}\ T^{10}G^{11}T^{12}$)-$3'$

$3'$-d($C^{24}A^{23}C^{22}G^{21}C^{20}G^{19}C^{18}A^{17}A^{16}A^{15}C^{14}A^{13}$)-$5'$

*Scheme 1. **A**. Interconversion of the cis-(5R,6S) and trans-(5R,6R) Tg lesions. When the 5R-Tg isomer is paired opposite dA in the 5'-GTgG-3' sequence a 7:3 cis-(5R,6S): trans-(5R,6R) mixture is present at equilibrium, in slow exchange on the NMR time scale (28). **B**. Oligodeoxynucleotide duplex containing the Tg•A pair used for NMR studies, indicating the numbering of the nucleotides. X^6 is the 5R-Tg lesion. **C**. Oligodeoxynucleotide duplex containing the Tg•G pair used for NMR studies, indicating the numbering of the nucleotides. X^6 is the 5R-Tg lesion.*

The 5*R*-Tg lesion is a substrate for base excision repair. This is mediated by at least two DNA *N*-glycosylase/AP lyases that are influenced by the diastereoisomer of Tg, the *cis–trans* epimerization of each diastereoisomer, and the identity of the complementary purine (*22*). The 5*R*-Tg lesion is also repaired by nucleotide excision repair (NER), although the effects of the *cis–trans* epimerization of each diastereoisomer of Tg with regard to NER have not been characterized. Both randomly introduced 5*R*-Tg and abasic sites are substrates for the *Escherichia coli* UvrABC proteins (*23, 24*). Tg is also excised *in vitro* by human NER (*25*). Earlier studies examined the structure of the 5*R*-Tg lesion in the 5′-ATgA-3′ sequence, paired opposite dA; it was concluded that Tg was partially extrahelical (*26*). It was also reported that the structure of 5*R*-Tg placed opposite dA in the 5′-GTgC-3′ sequence was disordered (*27*). These studies (*26*, *27*) did not address *cis-trans* epimerization (*5–7*) of the lesion. Consequently, we have examined the chemistry and structure of the 5*R*-Tg lesion in the 5′-GTgG-3′ sequence context, when paired opposite either dA or dG (*28–30*).

The 5′-d(GTGCGTgGTTTGT)-3′ Oligodeoxynucleotide

The dodecamer 5′-d(GTGCGTgGTTTGT)-3′, Tg = 5*R* Tg, was synthesized; its identity was confirmed by mass spectrometry (*31*). It eluted from reverse phase HPLC as a chromatographically separable species. NMR data for the duplexes containing the $Tg^{6}{\bullet}G^{19}$ or $Tg^{6}{\bullet}A^{19}$ base pairs were collected immediately upon sample preparation and subsequently repeated after 4 weeks; no changes in the spectra were observed, suggesting that the samples had achieved equilibrium with respect to *cis-trans* epimerization of the 5*R*-Tg lesion.

Recognition and Incision of the 5*R*-Tg Substrate by *Escherichia coli* UvrABC Proteins

A 51 bp substrate containing 5*R*-Tg was utilized. For comparison, a 50-bp substrate containing an AAF adduct (AAF-50bp) recognized and incised by *E. coli* UvrABC (*32*), was used. A 50-bp non-damaged substrate (ND-50bp) was used as a control. Dissociation constants measured from gel mobility shift assays showed that the ratio of binding affinities of UvrA to the AAF- vs. 5*R*-Tg substrate was 2.4--UvrA exhibited a K_d of 24 ± 2 nM for the Tg-50bp substrate, and it exhibited a K_d of 10 ± 1 nM for the AAF-51bp substrate (*30*), the latter value being in agreement with earlier reports (*32*). The UvrABC incision rate of the 5*R*-Tg substrate was greater than that of the helix-distorting cross-linked tandem G(8,5-Me)T lesion (*32*). Thus, in this sequence 5*R*-Tg was a substrate for the *E. coli* UvrABC proteins. However, the 5*R*-Tg substrate was incised 1.7x less efficiently than was the AAF substrate (0.48 ± 0.04 fmol/min vs. 0.80 ± 0.02 fmol/min) (*30*).

Binding of Human NER Proteins with the 5*R*-Tg Substrate

When the binding of XPA and XPC/HR23B to the 5*R*-Tg damaged duplex was compared with the AAF-damaged duplex (*30*), at low concentrations XPA bound to AAF-50bp with a greater affinity than to Tg-51bp, but at concentrations greater than 50 nM, XPA bound to both substrates with comparable affinity. The dissociation constants estimated for the 5*R*-Tg and AAF substrates were similar, 48 ± 4 nM and 44 ± 6 nM, respectively. XPA is believed to be recruited to DNA damage sites after XPC and TFIIH. However, XPA can recognize DNA damage by itself. Whether this *in vitro* activity of XPA has a role in NER or in cellular DNA damage responses is unclear. However, XPC/HR23B bound to Tg-51bp with a

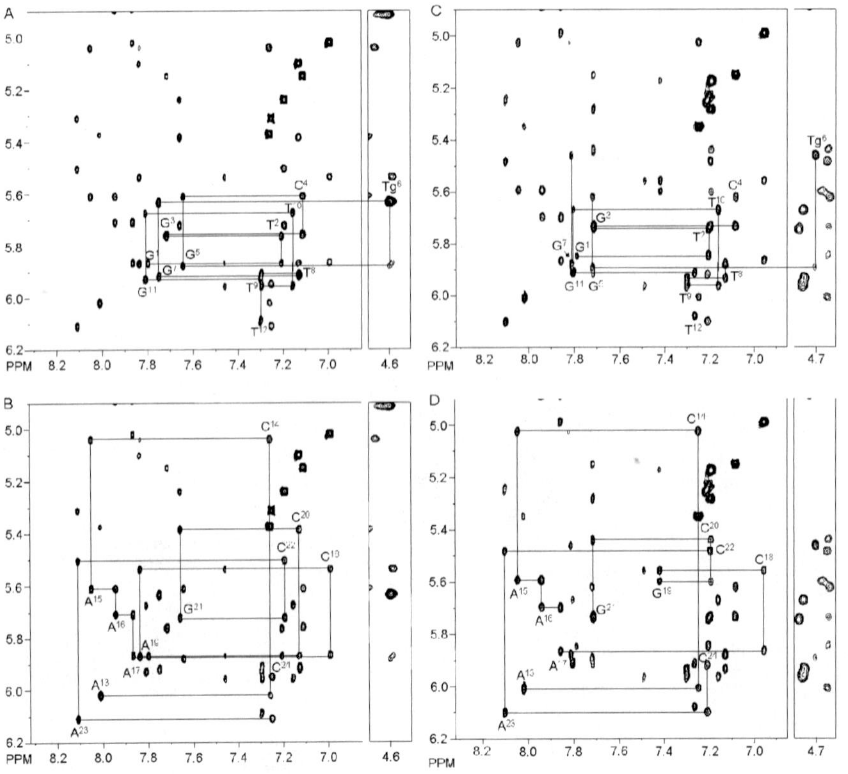

Figure 1. Sequential NOESY assignments of purine H8 and pyrimidine H6 protons to deoxyribose H1′ protons, for the duplexes containing either the X^6•A^{19} or X^6•G^{19} base pairs. ***A****. The modified strand of the duplex containing the X^6•A^{19} base pair.* ***B****. The complementary strand of the duplex containing the X^6•A^{19} base pair.* ***C****. The modified strand of the duplex containing the X^6•G^{19} base pair.* ***D****. The complementary strand of the duplex containing the X^6•G^{19} base pair. The intra-nucleotide aromatic to H1′ cross peaks are labeled. The data were collected at 800 MHz at 250 ms mixing time at 30 °C. Reprinted with permission from Brown, K.L., et al. (28). Copyright 2008 American Chemical Society.*

greater affinity than to AAF-50bp (*30*). These data are consistent with the role of XPC/HR23B as the major DNA damage recognition factor in NER (*33*, *34*).

NMR Spectroscopy

(a) Non-Exchangeable DNA Protons

NMR resonances were assigned using standard strategies (Figure 1) (*35*, *36*). For the duplex containing the X^6•A^{19} pair (*28*), there was no break in the sequential NOE connectivity for the modified strand. The G^5 H1′→Tg H6 NOE was observed, as was the Tg H6→Tg H1′ NOE. There was also no break in connectivity for the complementary strand. The duplex containing the X^6•G^{19} pair (*28*) also showed no break in connectivity for the modified strand. Again, the

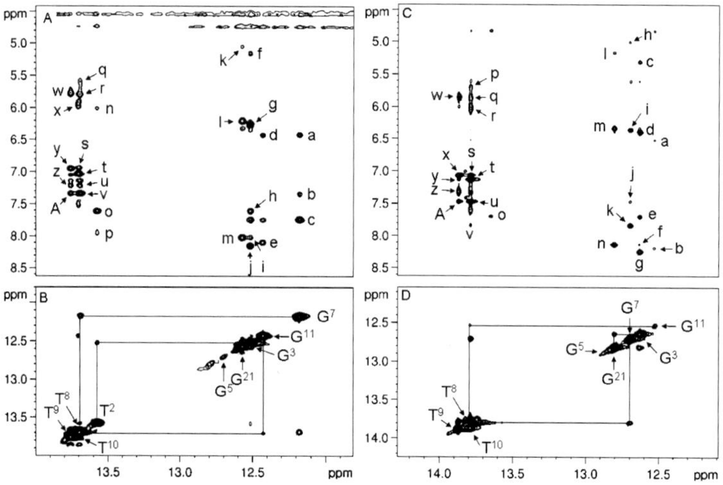

*Figure 2. **A**. Expanded plot showing NOEs from the imino protons to amino protons for the duplex containing the $X^{6}{\bullet}A^{19}$ base pair. The cross-peaks are assigned as a, G^{7} $N^{2}H2{\rightarrow}G^{7}$ N1H; b, A^{17} $H2{\rightarrow}G^{7}$ N1H; c, C^{18} $N^{4}H2{\rightarrow}G^{7}$ N1H1; d, G^{11} $N^{2}H2{\rightarrow}G^{11}$ N1H; e, C^{14} $N^{4}H2{\rightarrow}G^{11}$ N1H; f, C^{22} $H5{\rightarrow}G^{3}$ N1H; g, G^{3} $N^{2}H2{\rightarrow}G^{3}$ N1H; h, A^{23} $H2{\rightarrow}G^{3}$ N1H; i, C^{4} $N^{4}H2{\rightarrow}G^{3}$ N1H; j, C^{22} $N^{4}H2{\rightarrow}G^{3}$ N1H; k, C^{4} $H5{\rightarrow}G^{21}$ N1H; l, G^{21} $N^{2}H2{\rightarrow}G^{21}$ N1H; m, C^{4} $N^{4}H2{\rightarrow}G^{21}$ N1H; n, A^{19} $N^{6}H2{\rightarrow}T^{2}$ N3H; o, A^{23} $H2{\rightarrow}T^{2}$ N3H; p, A^{23} $N^{6}H1{\rightarrow}T^{2}$ N3H; q, A^{17} $N^{6}H2{\rightarrow}T^{8}$ N3H; r, A^{16} $N^{6}H2{\rightarrow}T^{8}$ N3H; s, A^{16} $H2{\rightarrow}T^{10}$ N3H; t, A^{15} $H2{\rightarrow}T^{10}$ N3H; u, A^{16} $N^{6}H1{\rightarrow}T^{8}$ N3H; v, A^{17} $H2{\rightarrow}T^{8}$ N3H; w, A^{16} $N^{6}H2{\rightarrow}T^{9}$ N3H; x, A^{15} $N^{6}H2{\rightarrow}T^{10}$ N3H; y, A^{16} $H2{\rightarrow}T^{9}$ N3H; z. A^{16} $N^{6}H1{\rightarrow}T^{9}$ N3H; A, A^{17} $H2{\rightarrow}T^{9}$ N3H. **B**. Expanded plot showing the sequential NOE connectivity for the imino protons. **C**. Expanded plot showing NOEs from the imino protons to amino protons for the duplex containing the $X^{6}{\bullet}G^{19}$ base pair. The cross-peaks are assigned as follows: a, G^{11} $N^{2}H2{\rightarrow}G^{11}$ N1H; b, C^{14} $N^{4}H2{\rightarrow}G^{11}$ N1H; c, C^{22} $H5{\rightarrow}G^{3}$ N1H; d, G^{3} $N^{2}H2{\rightarrow}G^{3}$ N1H; e, A^{23} $H2{\rightarrow}G^{3}$ N1H; f, C^{4} $N^{4}H2{\rightarrow}G^{3}$ N1H; g, C^{22} $N^{4}H2{\rightarrow}G^{3}$ N1H; h, C^{18} $H5{\rightarrow}G^{7}$ N1H; i, G^{7} $N^{2}H2{\rightarrow}G^{7}$ N1H; j, A^{17} $H2{\rightarrow}G^{7}$ N1H; k, C^{18} $N^{4}H2{\rightarrow}G^{7}$ N1H; l, C^{4} $H5{\rightarrow}G^{21}$ N1H; m, G^{21} $N^{2}H2{\rightarrow}G^{21}$ N1H; n, C^{4} $N^{4}H2{\rightarrow}G^{21}$ N1H; o, A^{23} $H2{\rightarrow}T^{2}$ N3H; p, A^{17} $N^{6}H2{\rightarrow}T^{8}$ N3H; q, A^{16} $N^{6}H2{\rightarrow}T^{8}$ N3H; r, A^{15} $N^{6}H2{\rightarrow}T^{10}$ N3H; s, A^{16} $H2{\rightarrow}T^{10}$ N3H; t, A^{15} $H2{\rightarrow}T^{10}$ N3H; u, A^{17} $H2{\rightarrow}T^{8}$ N3H; v, C^{18} $N^{4}H2{\rightarrow}T^{8}$ N3H; w, A^{16} $N^{6}H2{\rightarrow}T^{9}$ N3H; x, A^{16} $H2{\rightarrow}T^{9}$ N3H; y, A^{15} $H2{\rightarrow}T^{9}$ N3H; z, A^{16} $N^{6}H1{\rightarrow}T^{9}$ N3H; A, A^{17} $H2{\rightarrow}T^{9}$ N3H. **D**. Expanded plot showing the sequential NOE connectivity for the imino protons of the duplex containing the $X^{6}{\bullet}G^{19}$ base pair. The data were collected at 800 MHz at 250 ms mixing time at 7 °C. Reprinted with permission from Brown, K.L., et al. (28). Copyright 2008 American Chemical Society.*

G^{5} H1′→Tg H6 NOE was observed, as was the Tg H6→Tg H1′ NOE. There was also no break in connectivity for the complementary strand. For both duplexes, assignments for the deoxyribose protons were made unequivocally with the exception of several of the H4′ protons, and the stereotopic assignments of the H5′ and H5" protons.

(b) Exchangeable DNA Protons

NMR spectra showing the imino protons of the two duplexes are shown in Figure 2. The Tg N3H imino resonance was not identified for either duplex. This was attributed to rapid exchange with solvent. The assignments of the remaining Watson-Crick hydrogen-bonded imino and amino protons of the two modified oligodeoxynucleotides were made using standard methods (*37*). The spectra were similar for both of the 5*R*-Tg-modified duplexes (Figure 2) (*28*). In both instances, the G^5 N1H imino resonance was broad at 5 °C and disappeared when the temperature was increased to 15 °C, indicating that the presence of Tg influenced Watson-Crick hydrogen bonding at the 5′ neighbor $G^5{\bullet}C^{20}$ base pair. In contrast, for the unmodified sample, the G^5 N1H imino resonance was sharp and was observed at temperatures as high as 40 °C. For both modified duplexes, there was no cross peak between the broad G^5 N1H resonance and G^{21} N1H, located at base pair $C^4{\bullet}G^{21}$. This was attributed to its exchange with solvent. The imino resonances for base pairs $T^2{\bullet}A^{23}$, $G^3{\bullet}C^{22}$, $C^4{\bullet}G^{21}$, $G^7{\bullet}C^{18}$, $T^8{\bullet}A^{17}$, $T^9{\bullet}A^{16}$, $T^{10}{\bullet}A^{15}$, and $G^{11}{\bullet}C^{14}$ were observed. The imino resonances for the terminal base pairs $G^1{\bullet}C^{24}$ and $T^{12}{\bullet}A^{13}$ were not observed, attributed to exchange broadening with water.

(c) Tg Protons

For the duplex containing the $X^6{\bullet}A^{19}$ pair NOESY data showed two cross-peaks arising from dipolar couplings between Tg CH_3 and Tg H6 (Figure 3) (*28*). These were attributed to exchange between two chemical species, involving the Tg CH_3 and Tg H6 protons. The volumes of the two exchange cross peaks were consistent at multiple NOE mixing times. Integration of the two Tg CH_3 resonances indicated that the two species were present at an equilibrium ratio of 7:3. For the major species, the Tg CH_3 protons exhibited a chemical shift of 0.49 ppm, while the Tg H6 proton resonated at 4.58 ppm. These chemical shifts were consistent with values reported in earlier NMR studies (*26*). Twenty-three NOE cross peaks were assigned between Tg CH_3 and H6 in the major species and DNA (seven for Tg H6 and sixteen for Tg CH_3). The resonances for the Tg CH_3 and H6 protons of the minor species were significantly downfield relative to those from the major species, located at 1.24 ppm and 4.91 ppm, respectively. The Tg CH_3 resonance for the minor species was overlapped with the T^2 CH_3 resonance. For the minor species, there was only one NOE cross peak observed to a DNA proton, observed between Tg H6 and Tg H2′. A single set of resonances was observed for the G^5 and G^7 DNA protons, indicating that the two species observed for Tg did not extend to the neighboring nucleotides. For the duplex containing the $X^6{\bullet}G^{19}$ pair (*28*) analysis of NOESY data obtained at multiple mixing times did not exhibit chemical exchange cross peaks for either the Tg CH_3 and Tg H6 protons (Figure 3). This indicated that only one chemical species was significantly populated. The Tg CH_3 protons exhibited a chemical shift of 0.91 ppm, while the Tg H6 proton resonated at 4.70 ppm. Twenty NOE cross peaks were assigned between Tg CH_3 and H6 and DNA (nine for Tg H6 and eleven for Tg CH_3).

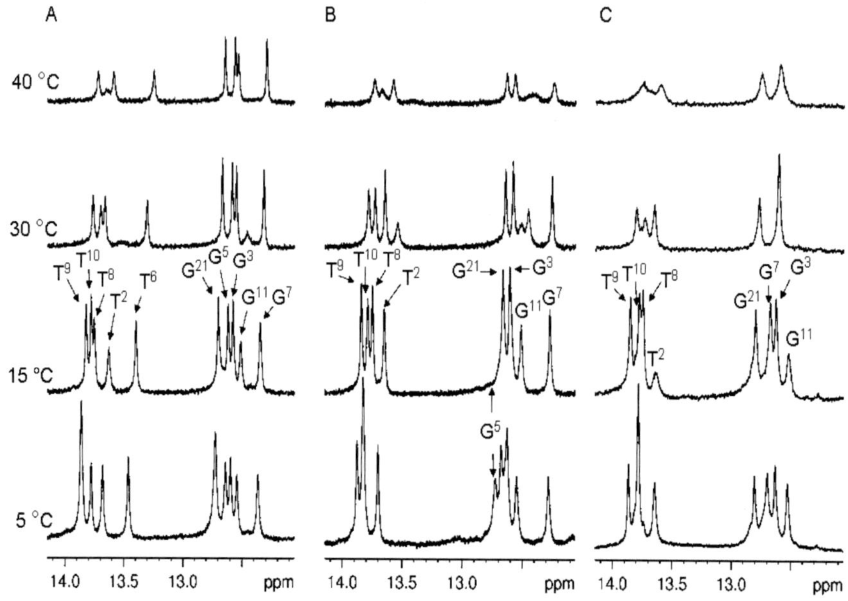

Figure 3. Temperature dependent analysis of imino protons of the duplexes containing (A) the $T^6{\bullet}A^{19}$ base pair, (B) the $X^6{\bullet}A^{19}$ base pair, and (C) the $X^6{\bullet}G^{19}$ base pairs, as monitored by ^1H NMR. Reprinted with permission from Brown, K.L., et al. (28). Copyright 2008 American Chemical Society.

Equilibrium between the *cis*-5*R*,6*S* and *trans*-5*R*,6*R* Epimers in Duplex Oligodeoxynucleotides Depends Upon the Purine Opposite Tg

The two Tg species observed for the duplex containing the $X^6{\bullet}A^{19}$ base pair were assigned as arising from slow exchange between the *cis*-(5*R*,6*S*) and *trans*-(5*R*,6*R*) epimers (*28*). Seven NOESY cross peaks were observed between the Tg H6 resonance of the major species and surrounding protons. Their intensities at mixing times of 80 and 250 ms were compared with the corresponding distances predicted for each epimer on the basis of molecular modeling. The spectral overlap of Tg H3′ and Tg H6 resonances in the major species hindered the assessment of the Tg H3′→Tg H1′ and Tg H6→Tg H1′ cross peaks. For the *cis*-(5*R*,6*S*) configuration, Tg H6 and Tg CH_3 are spatially proximate, which resulted in a strong Tg H6→Tg CH_3 NOE even at the mixing time of 150 ms (Figure 4). Likewise, the G^5 H1′→Tg H6 and G^5 H8→Tg H6 NOEs were diagnostic of the *cis*-5*R*,6*S* configuration. On this basis, the major species, present at ~ 70% population, was assigned as the *cis* epimer. The configuration of the single species present in the duplex containing the $X^6{\bullet}G^{19}$ base pair was determined to be *cis* by the same approach. For the duplex containing the $X^6{\bullet}G^{19}$ base pair the Tg H3′ and Tg H6 resonances were separated at 4.53 and 4.70 ppm, respectively.

For the duplex containing the $Tg^6{\bullet}A^{19}$ pair, the 5*R*-Tg adduct exists as an 70%:30% mixture of *cis*-(5*R*,6*S*) and *trans*-(5*R*,6*R*) epimers (*28*). This is

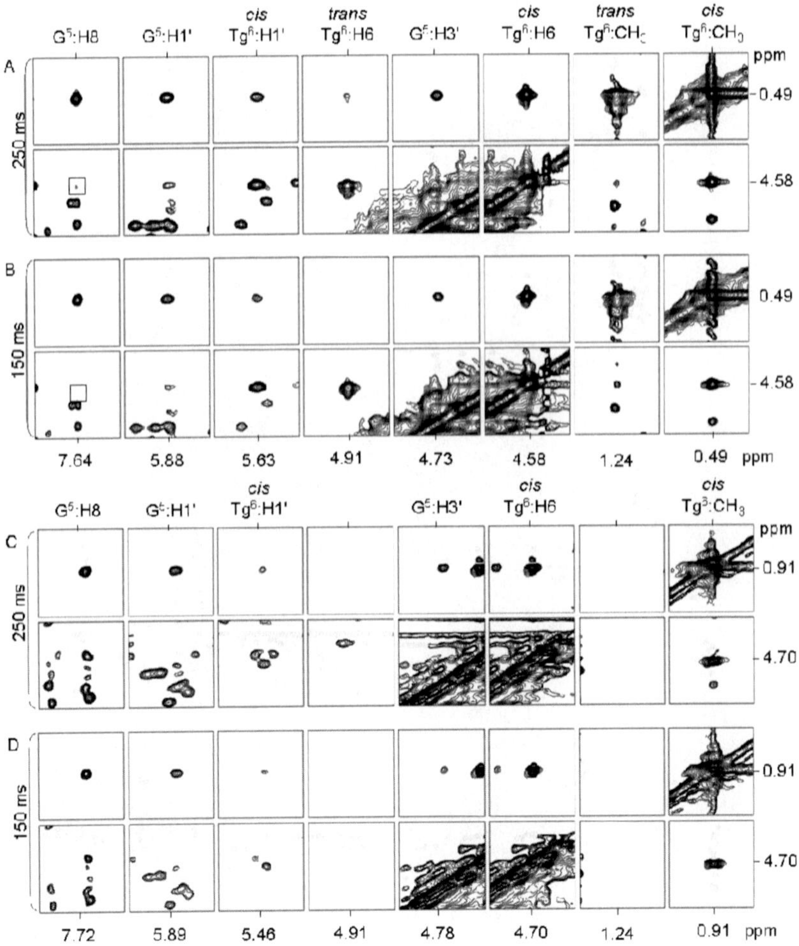

*Figure 4. **A**. NOESY data collected for the duplex containing the X^6•A^{19} base pair at a NOE mixing time of 250 ms. **B**. NOESY data collected for the duplex containing the X^6•A^{19} base pair at a NOE mixing time of 150 ms. **C**. NOESY data collected for the duplex containing the X^6•G^{19} base pair at a NOE mixing time of 250 ms. **D**. NOESY data collected for the duplex containing the X^6•G^{19} base pair at a NOE mixing time of 150 ms. The spectra were collected at 800 MHz at 30 °C. Reprinted with permission from Brown, K.L., et al. (28). Copyright 2008 American Chemical Society.*

comparable to the 87%:13% *cis:trans* mixture of epimers at the nucleoside level (*6*). In contrast, for the duplex containing the Tg^6•G^{19} pair, Tg^6 exists predominantly as the *cis*-(5*R*,6*S*) epimer (*29*). Thus, depending upon the identity of the complementary nucleotide, significant levels of the *trans*-(5*R*,6*R*) epimer may be present in DNA, and the 5*R*-Tg lesion should be considered to exist in DNA as a mixture of the two epimers.

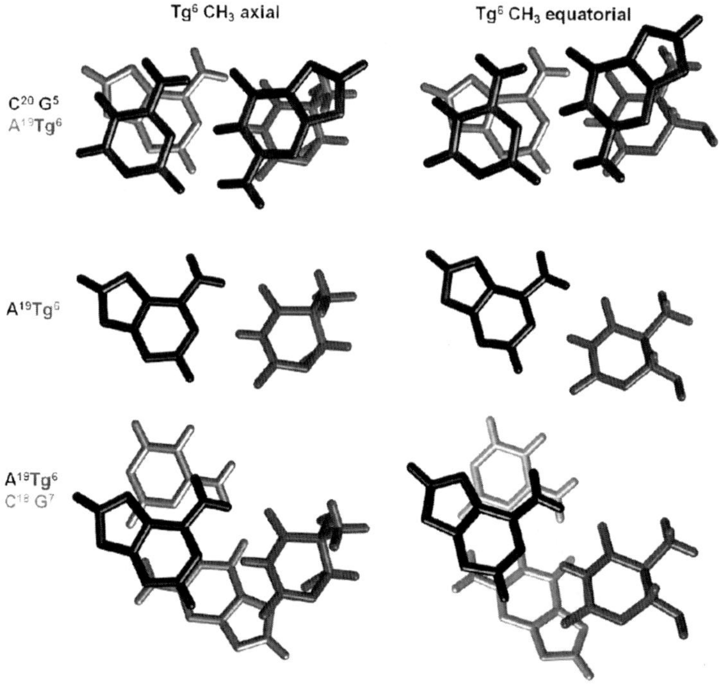

Figure 5. Base pair stacking interactions of the cis-(5R,6S) Tg lesion in the Tg•A pair. Comparison of stacking interactions in which Tg CH_3 is in the axial (PDB ID 2KH5) or equatorial (PDB ID 2KH6) conformations. The top panel shows the G^5•C^{20} base pair (black) stacked above the X^6•A^{19} base pair (Tg is colored dark grey (red online) and A^{19} is colored light grey). The center panel shows the orientation of the X^6 lesion (dark grey (red online)) with respect the complementary nucleotide A^{19} (black). The bottom panel shows the X^6•A^{19} base pair (Tg is colored dark grey (red online) and A^{19} is colored black) stacked above the G^7•C^{18} base pair (light grey). Reprinted with permission from Brown, K.L., et al. (30). Copyright 2010 Oxford University Press.

Structural Refinement

For both 5*R*-Tg modified duplexes thirty starting structures were generated, of which half had the Tg CH_3 group in the axial conformation and half had the Tg CH_3 group in the equatorial conformation (*29*, *30*). Both of the duplexes were subjected to simulated annealing rMD calculations, using the generalized Born approach (*38*, *39*) for modeling solvent. For each of the duplexes the structures emergent from simulated annealing calculations typically exhibited Tg CH_3 in the axial conformation, although several structures were observed in which Tg CH_3 was in the equatorial conformation. After checking that the structures emergent from the simulated annealing protocol were in agreement with the NMR data (*40*, *41*), for each of the duplexes a representative structure with Tg CH_3 in the axial conformation was placed into a truncated octahedron TIP3P water box, and subjected to 10 ns of isothermal rMD calculations at 300 K.

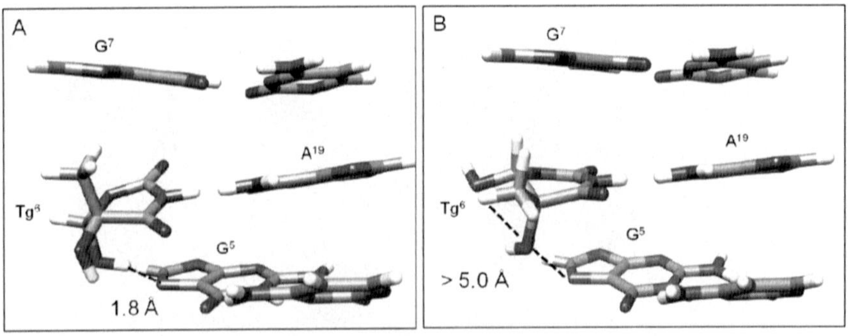

Figure 6. The cis-(5R,6S)Tg lesion at the X^6•A^{19} base pair as viewed from the major groove showing potential hydrogen bonding interactions as predicted from analyses of rMD trajectories. ***A****. The Tg OH6 formed a hydrogen bond with G^5 N7 when Tg CH_3 was in the axial conformation (PDB ID 2KH5).* ***B****. When Tg CH_3 was in the equatorial conformation Tg OH6 did not hydrogen bond with G^5 N7, however, improved hydrogen bonding was observed with Tg OH5 (PDB ID 2KH6). Reprinted with permission from Brown, K.L., et al. (30). Copyright 2010 Oxford University Press.*

For the duplex containing the X^6•A^{19} base pair (*30*), with the Tg CH_3 group in the axial conformation, a 42% occupancy of the Tg HO6→G^7 N7 hydrogen bond was observed. This induced altered propeller twist at the lesion site. In contrast, the distance between Tg HO5 and G^7 N7 was 4.5 Å. When Tg CH_3 was oriented in the equatorial conformation, Tg OH5 shifted to the axial conformation, which allowed for hydrogen bond formation between Tg OH5 and G^7 N7 with an occupancy of 11%. Structural ensembles representing both axial and equatorial conformations of Tg CH_3 were extracted and complete relaxation matrix analyses (*40, 41*) were performed. While both ensembles satisfied the NOESY data, at the lesion site, the agreement between the structures with the axial conformation and the data was modestly improved.

For the duplex containing the X^6•G^{19} base pair (*29*), the Tg HO6→G^7 N7 hydrogen bond exhibited occupancy of only 2.5 % when Tg CH_3 was axial. This altered propeller twist at the lesion site and resulted in a 6 Å Tg HO5→G^7 N7 distance, which prevented hydrogen bond formation. When Tg CH_3 oriented in the equatorial conformation, Tg OH5 became axial and exhibited improved geometry for hydrogen bond formation with G^7 N7. The analysis indicated 16% occupancy for this motif. Ensembles of structures were extracted in which the Tg CH_3 was in the axial and in the equatorial conformations were extracted; complete relaxation matrix analyses (*40*, *41*) were performed. The calculated sixth root residual R_1^X values indicated agreement between both ensembles and the NOESY data, although again, agreement between the structures with the axial conformation and the data was modestly improved.

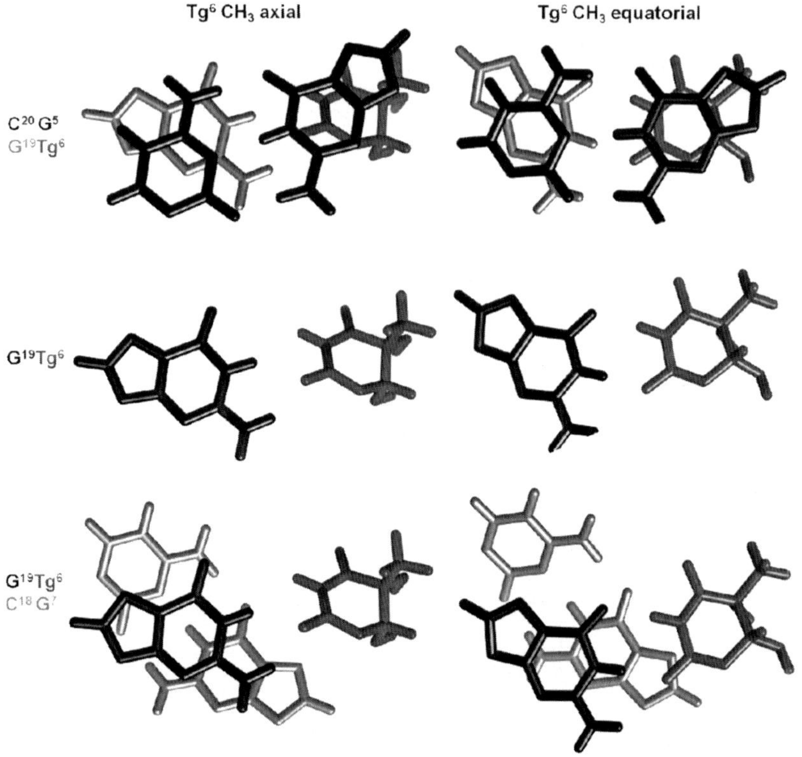

Figure 7. The wobble orientation of Tg⁶ in the Tg⁶•G¹⁹ mismatch pair does not favor the formation of stabilizing intra-strand hydrogen bonds between the hydroxyl groups of the cis-5R,6S Tg⁶ lesion and the imidazole ring N7 atom of the 5′-neighbor purine, as determined from analyses of rMD trajectories in explicit solvent. ***A****. When Tg⁶ CH_3 is in the axial conformation, the distance between the C6 OH group and G7 N7 is > 5 Å.* ***B****. When Tg⁶ CH_3 is in the equatorial conformation, the distance between the C6 OH group and G7 N7 is > 4 Å. Reprinted with permission from Brown, K.L., et al. (29). Copyright 2009 American Chemical Society.*

Structure of the X^6•A^{19} Pair in the 5′-GTgG-3′ Sequence

The *cis*-(5*R*,6*S*) epimer exhibits a Watson-Crick type alignment in which Tg O^4 is proximate to the exocyclic amine of A^{19}, and Tg N3H is proximate to A^{19} N1 (Figure 5) (*30*). Both the Tg amine and A^{19} N^6 amine resonances protons undergo increased exchange with solvent (*28*). In the 3′-direction, stacking between the *cis*-(5*R*,6*S*) Tg and base pair G^7•C^{18} is disrupted (Figure 6). As compared to a Watson-Crick T•A base pair the *cis*-(5*R*,6*S*) Tg lesion is more exposed to solvent. This agrees with findings in the 5′-AXA-3′ sequence when the 5*R*-Tg lesion was placed opposite dA (*26*). The Tg lesion remains stacked into the duplex and it is not flipped into the major groove (Figure 6). It has been predicted that the Tg•A pair should be stabilized by an intra-strand hydrogen bond between the

Tg OH6 and the N7 position of a 3′ purine (*18*). The rMD trajectories suggest that this hydrogen bond is present when Tg CH_3 is in the axial conformation; its occupancy is 46%. It induces propeller twist at the lesion site, which alleviates steric interactions between the CH_3 group and the 5′-neighbor guanine. This is reflected in the perturbation of the glycosyl bond at Tg. The increase in longitudinal relaxation for the Tg CH_3 protons (*28*) is consistent with the axial conformation the Tg CH_3 group, which orients it toward G^5. It exhibits NOEs to G^5 protons; all are sources of longitudinal relaxation. The axial conformation of the Tg CH_3 group provides a modestly improved agreement with the NOE data, consistent with crystallographic data at the nucleoside level (*42*) and quantum mechanical calculations using the modified base (*18*). However, the equatorial conformation of the Tg CH_3 group is also observed in rMD trajectories. These indicate the potential for Tg OH5→G^7 N7 hydrogen bond formation when the Tg CH_3 group is in the equatorial conformation. This would alleviate steric clash between the Tg CH_3 group and the 5′-neighbor guanine. Though the trajectories predict only 11% occupancy of this hydrogen bond, its presence cannot be ruled out (*30*). However, epimerization to the *trans*-(5*R*,6*R*) Tg configuration, for which the equatorial conformation of the Tg CH_3 group is favored by 4 kcal/mol (*18*), probably represents the more favorable mechanism for alleviating steric strain (*28*). Notably, in the structure of the RB69 polymerase involving a template containing the *cis*-(5*R*,6*S*) Tg lesion and an incoming dATP, the Tg CH_3 group remains in the axial conformation, despite hindering stacking of the adjacent 5′-template guanine (*43*). Bolton and coworkers (*27*) reported that a disordered structure resulted when 5*R*-Tg was placed into a duplex containing the 5′-GTgC-3′ sequence. This suggests that formation of an intra-strand hydrogen between the Tg C6 OH and the N7 position of a 3′ purine (*18*) is important in stabilizing the *cis*-(5*R*,6*S*)-Tg lesion in duplex DNA.

Structure of the X^6•G^{19} Pair in the 5′-GTgG-3′ Sequence

The *cis*-(5*R*,6*S*) epimer stacks into the duplex and assumes the wobble position (*29*), similar to the unmodified T•G mismatch pair (*44*, *45*). The *cis*-(5*R*,6*S*) Tg epimer perturbs the 5′-neighbor base pair G^5•C^{20}. The imino resonance of base pair G^5•C^{20} broadens due to solvent exchange (Figure 2) and disappears from the spectrum ~ 35 °C lower as compared to the corresponding unmodified duplex (*28*). Tg shows poor stacking interactions with 3′-neighbor base pair G^7•C^{20}. As seen for the X^6•A^{19} pair, the axial conformation orients the Tg CH_3 group in the 5′ direction, facing toward base pair G^5•C^{20}. In this case, steric interaction between the Tg CH_3 and the 5′ neighboring G^5 is reduced by wobble pairing (*28*). However, the observation that the rMD trajectories yield structures in which the Tg CH_3 is in either the axial or equatorial conformations suggests that both conformers of the Tg ring may be populated. However, for the X^6•G^{19} pair, the modest occupancies of the Tg OH6 to G^7 N7 hydrogen bond associated with the axial conformation of the Tg CH_3 group, and the Tg OH5 to G^7 N7 hydrogen bond associated with the equatorial conformation of the Tg CH_3 group, suggest that intra-strand hydrogen bond formation between the Tg

C6 hydroxyl and the N7 position of the 3′ purine (*18*) does not stabilize either the axial or equatorial conformations of the Tg base (Figure 7). This is consistent with an increase in transverse relaxation rate for Tg CH_3 as compared to unmodified thymine CH_3 protons, which is attributed to puckering of the Tg six-member ring between axial and equatorial conformations of the Tg CH_3 group. However, increased backbone and deoxyribose disorder in tandem with the puckering of the Tg ring cannot be excluded (*44*, *46*, *47*). Tg is more exposed to solvent as compared to a Watson-Crick T•A base pair. This is in agreement with the findings of Kung and Bolton (*26*) in the 5′-AXA-3′ sequence when the 5*R*-Tg lesion was placed opposite dA. It is consistent with the wobble orientation of Tg, which shifts the modified base toward the major groove. However, the Tg lesion remains stacked into the duplex and it is not flipped into the major groove (Figure 8).

Biological Implications

The 5*R*-Tg lesion is a substrate for nucleotide excision repair in *E. coli* (*23*), and it is excised *in vitro* by human NER proteins (*25*). DNA containing dihydrothymine, a lesion with a similar structure to thymine glycol, but which cannot undergo epimerization between *cis* and *trans* epimers, is not incised (*24*). However, the effect of *cis-trans* epimerization and the position of this equilibrium with respect to the complementary purine in modulating NER remain to be determined. The binding of the AAF-dG damaged DNA by XPC/HR23B has been attributed to the inability of the damaged dG base to pair efficiently with cytosine (*48*), and an emerging consensus posits that disruption of normal base pairing and the resulting destabilization of the helix, rather than the recognition of helical distortion by bulky lesions, governs the affinity of XPC/HR23B binding (*49*). From studies of the yeast XPC orthologue Rad4 bound to DNA containing a cyclobutane pyrimidine dimer, Min and Pavletich (*50*) concluded that damage recognized by Rad4 destabilizes the helix and facilitates the flipping out of two base pairs by the protein. Our structural studies demonstrate that 5*R*-Tg paired opposite dA in the 5′-GTgG-3′ sequence perturbs the 5′-neighbor base pair $G^5{\bullet}C^{20}$, in addition to the damaged base pair $Tg^6{\bullet}A^{19}$ (*28*). Similarly, based upon MD simulations of the Dickerson-Drew 5*R*-Tg-modified dodecamer, Miller et al. (*51*) and Miaskiewicz et al. (*52*) concluded that Tg weakened Watson-Crick hydrogen bonds of the 5′-neighbor base pair. This may facilitate flipping both base pairs out of the helix, enabling XPC/HR23B to bind the 5*R*-Tg lesion.

In *E. coli*, base excision repair of Tg is initiated by endonuclease III (Nth) (*53*) and endonuclease VIII (Nei) (*54*). Yeast (*55*), mammalian (*56*, *57*), and human orthologs (*58*–*60*) of Nth have been characterized. Likewise, human orthologs of Nei have been characterized (*61*, *62*). The human hNTH1 exhibits a 13:1 preference for excising the 5*R* vs. the 5*S* epimers, whereas hNEIL1 (*61*, *63*) shows a 1.5:1 preference for excising the 5*R* vs. the 5*S* epimers (*64*). Similar observations have been made for prokaryotic, yeast, and murine glycosylases (*65*). The differential structural consequences of *cis-trans* epimerization of the 5*R*-Tg lesion in DNA reported herein are significant in light of observations by Ocampa-Hafalla et al. (*22*), showing that the base excision repair of 5*R*-Tg

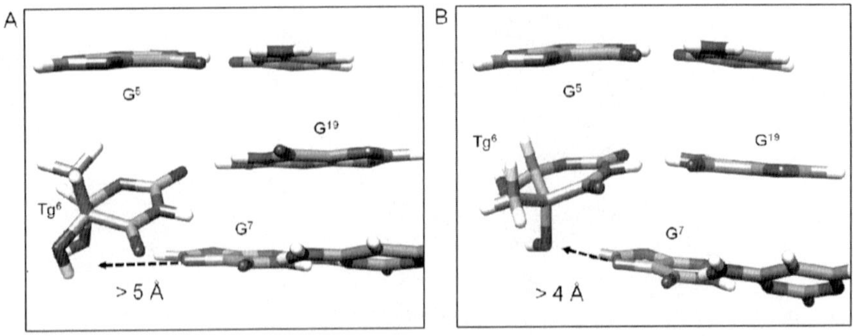

Figure 8. Base pair stacking interactions determined from structural refinements using rMD calculations restrained by NMR-derived distance and torsion angles, for the cis-5R,6S-Tg-adducted duplex containing the $Tg^6{\bullet}G^{19}$ mismatch pair. The left panel shows stacking interactions when Tg^6 CH_3 is in the axial conformation (PDB ID: 2KH7). The right panel shows stacking interactions when Tg^6 CH_3 is in the equatorial conformation (PDB ID: 2KH8). In both instances, the Tg^6 base shifts toward the major groove, into a wobble orientation with G^{19} in the complementary strand, but both Tg^6 and G^{19} remain stacked into the duplex. Reprinted with permission from Brown, K.L., et al. (29). Copyright 2009 American Chemical Society.

by N-glycosylases/AP lyases is modulated by *cis-trans* epimerization and that repair of 5*R*-Tg by hNEIL1 depends upon the opposing base, with Tg•G pairs being excised more rapidly than Tg•A pairs. Computational studies suggest that substrates for hNEIL1 possess in common a pyrimidine-like ring and hydrogen bond donor-acceptor properties, allowing them to be accommodated within the enzyme's binding pocket (*66*). In the Tg•G pair the wobble orientation of the *cis*-(5*R*,6*S*)-Tg^6 base shifts it toward the major groove, reflected in an increased solvent accessible surface and reduced barrier for breathing of the lesion (*29*). While the Tg•A structure also shows increased solvent accessible surface area, it differs from the Tg•G pair in that a hydrogen bond between Tg OH6→G^7 N7 is facilitated; this may be sufficient to increase the barrier toward base-flipping of the Tg lesion into the active site pocket of the glycosylase, hindering repair. Thus, hNeil1's ability to excise 5*R*-Tg more efficiently in a Tg•G pair compared to a Tg•A pair (*22*) may be related to the exclusive *cis* epimerization in the former pair.

Summary

Our structural studies indicate that the *cis-trans* epimerization of the 5*R*-Tg lesion may play an important role in modulating both base excision repair and nucleotide excision repair of the lesion. In the 5′-GTG-3′ sequence, the equilibrium between the *cis*-5*R*,6*S*-and *trans*-5*R*,6*R*- epimers of the 5*R*-Tg lesion depends upon the identity of the purine in the complementary strand. For the duplex containing the Tg•A pair at 30 °C, the *cis*:*trans* ratio is 70%:30%. In contrast, for the duplex containing the Tg•G pair, equilibrium favors the *cis* epimer. In this sequence, 5*R*-Tg paired with dA is a good substrate for the NER proteins from *E. Coli*, corroborating earlier reports (*23–25*). The human NER proteins XPA and XPC/HR23B bind more strongly to 5*R*-Tg paired with dA in this sequence, as compared to the C8-dG adduct of AAF. The *cis*-(5*R*,6*S*) Tg epimer destabilizes the modified base pair as well as the 5′-neighbor G·C base pair, which may facilitate binding by XPC/HR23B in nucleotide excision repair. The ability of the hNeil1 glycosylase to excise 5*R*-Tg more efficiently in a Tg•G pair compared to a Tg•A pair (*22*) may be related to the observation that for the Tg•A pair a hydrogen bond between Tg^6 OH6→G^7 N7 is facilitated. This may increase the energy barrier toward base-flipping of the Tg lesion into the active site pocket of the glycosylase, hindering repair.

Acknowledgments

This work was presented, in part, at the 236th National Meeting of the American Chemical Society, Philadelphia, PA. This work was funded by NIH grants CA-55678 (M.P.S.), ES-013324 (A.B), and CA-086927 (Y.Z.).

References

1. Teoule, R.; Bonicel, A.; Bert, C.; Cadet, J.; Polverelli, M. *Radiat. Res.* **1974**, *57*, 46–58.
2. Frenkel, K.; Goldstein, M. S.; Teebor, G. W. *Biochemistry* **1981**, *20*, 7566–7571.
3. Zuo, S.; Boorstein, R. J.; Teebor, G. W. *Nucleic Acids Res.* **1995**, *23*, 3239–3243.
4. Pfeifer, G. P. *Mutat. Res.* **2000**, *450*, 155–166.
5. Vaishnav, Y.; Holwitt, E.; Swenberg, C.; Lee, H. C.; Kan, L. S. *J. Biomol. Struct. Dyn.* **1991**, *8*, 935–951.
6. Lustig, M. J.; Cadet, J.; Boorstein, R. J.; Teebor, G. W. *Nucleic Acids Res.* **1992**, *20*, 4839–4845.
7. Wang, Y. *Chem. Res. Toxicol.* **2002**, *15*, 671–676.
8. Cathcart, R.; Schwiers, E.; Saul, R. L.; Ames, B. N. *Proc. Natl. Acad. Sci. U.S.A.* **1984**, *81*, 5633–5637.
9. Adelman, R.; Saul, R. L.; Ames, B. N. *Proc. Natl. Acad. Sci. U.S.A.* **1988**, *85*, 2706–2708.
10. Ide, H.; Kow, Y. W.; Wallace, S. S. *Nucleic Acids Res.* **1985**, *13*, 8035–8052.
11. Clark, J. M.; Beardsley, G. P. *Nucleic Acids Res.* **1986**, *14*, 737–749.
12. Achey, P. M.; Wright, C. F. *Radiat. Res.* **1983**, *93*, 609–612.
13. Moran, E.; Wallace, S. S. *Mutat. Res.* **1985**, *146*, 229–241.
14. Laspia, M. F.; Wallace, S. S. *J. Bacteriol.* **1988**, *170*, 3359–3366.
15. Hayes, R. C.; Petrullo, L. A.; Huang, H. M.; Wallace, S. S.; LeClerc, J. E. *J. Mol. Biol.* **1988**, *201*, 239–246.
16. Kow, Y. W.; Faundez, G.; Melamede, R. J.; Wallace, S. S. *Radiat. Res.* **1991**, *126*, 357–366.
17. Hayes, R. C.; LeClerc, J. E. *Nucleic Acids Res.* **1986**, *14*, 1045–1061.

18. Clark, J. M.; Pattabiraman, N.; Jarvis, W.; Beardsley, G. P. *Biochemistry* **1987**, *26*, 5404–5409.
19. McNulty, J. M.; Jerkovic, B.; Bolton, P. H.; Basu, A. K. *Chem. Res. Toxicol.* **1998**, *11*, 666–673.
20. Kusumoto, R.; Masutani, C.; Iwai, S.; Hanaoka, F. *Biochemistry* **2002**, *41*, 6090–6099.
21. Fischhaber, P. L.; Gerlach, V. L.; Feaver, W. J.; Hatahet, Z.; Wallace, S. S.; Friedberg, E. C. *J. Biol. Chem.* **2002**, *277*, 37604–37611.
22. Ocampo-Hafalla, M. T.; Altamirano, A.; Basu, A. K.; Chan, M. K.; Ocampo, J. E.; Cummings, A., Jr.; Boorstein, R. J.; Cunningham, R. P.; Teebor, G. W. *DNA Repair (Amst)* **2006**, *5*, 444–454.
23. Lin, J. J.; Sancar, A. *Biochemistry* **1989**, *28*, 7979–7984.
24. Kow, Y. W.; Wallace, S. S.; Van Houten, B. *Mutat Res.* **1990**, *235*, 147–56.
25. Reardon, J. T.; Bessho, T.; Kung, H. C.; Bolton, P. H.; Sancar, A. *Proc. Natl. Acad. Sci. U.S.A.* **1997**, *94*, 9463–9468.
26. Kung, H. C.; Bolton, P. H. *J. Biol. Chem.* **1997**, *272*, 9227–9236.
27. Kao, J. Y.; Goljer, I.; Phan, T. A.; Bolton, P. H. *J. Biol. Chem.* **1993**, *268*, 17787–17793.
28. Brown, K. L.; Adams, T.; Jasti, V. P.; Basu, A. K.; Stone, M. P. *J. Am. Chem. Soc.* **2008**, *129*, 11701–11710.
29. Brown, K. L.; Basu, A., K.; Stone, M. P. *Biochemistry* **2009**, *48*, 9722–9733.
30. Brown, K. L.; Roginskaya, M.; Zou, Y.; Altamirano, A.; Basu, A., K.; Stone, M. P. *Nucleic Acids Res.* **2010**, *38*, 428–440.
31. Iwai, S. *Nucleic Acids Symp. Ser.* **2000**, 121–122.
32. Yang, Z.; Colis, L. C.; Basu, A. K.; Zou, Y. *Chem. Res. Toxicol.* **2005**, *18*, 1339–1346.
33. Hey, T.; Lipps, G.; Sugasawa, K.; Iwai, S.; Hanaoka, F.; Krauss, G. *Biochemistry* **2002**, *41*, 6583–6587.
34. Riedl, T.; Hanaoka, F.; Egly, J. M. *EMBO J.* **2003**, *22*, 5293–5303.
35. Reid, B. R. *Q. Rev. Biophys.* **1987**, *20*, 1–34.
36. Patel, D. J.; Shapiro, L.; Hare, D. *Q. Rev. Biophys.* **1987**, *20*, 35–112.
37. Boelens, R.; Scheek, R. M.; Dijkstra, K.; Kaptein, R. *J. Magn. Res.* **1985**, *62*, 378–386.
38. Bashford, D.; Case, D. A. *Annu. Rev. Phys. Chem.* **2000**, *51*, 129–152.
39. Tsui, V.; Case, D. A. *Biopolymers* **2000**, *56*, 275–291.
40. Keepers, J. W.; James, T. L. *J. Magn. Reson.* **1984**, *57*, 404–426.
41. James, T. L. *Curr. Opin. Struct. Biol.* **1991**, *1*, 1042–1053.
42. Hruska, F. E.; Sebastian, R.; Grand, A.; Voituriez, L.; Cadet, J. *Can. J. Chem.* **1987**, *65*, 2618–2623.
43. Aller, P.; Rould, M. A.; Hogg, M.; Wallace, S. S.; Doublie, S. *Proc. Natl. Acad. Sci. U.S.A.* **2007**, *104*, 814–818.
44. Patel, D. J.; Kozlowski, S. A.; Marky, L. A.; Rice, J. A.; Broka, C.; Dallas, J.; Itakura, K.; Breslauer, K. J. *Biochemistry* **1982**, *21*, 437–444.
45. Hare, D.; Shapiro, L.; Patel, D. J. *Biochemistry* **1986**, *25*, 7445–7456.
46. Roongta, V. A.; Jones, C. R.; Gorenstein, D. G. *Biochemistry* **1990**, *29*, 5245–5258.
47. Moe, J. G.; Russu, I. M. *Biochemistry* **1992**, *31*, 8421–8428.

48. O'Handley, S. F.; Sanford, D. G.; Xu, R.; Lester, C. C.; Hingerty, B. E.; Broyde, S.; Krugh, T. R. *Biochemistry* **1993**, *32*, 2481–2497.
49. Sugasawa, K.; Shimizu, Y.; Iwai, S.; Hanaoka, F. *DNA Repair (Amst)* **2002**, *1*, 95–107.
50. Min, J. H.; Pavletich, N. P. *Nature* **2007**, *449*, 570–575.
51. Miller, J.; Miaskiewicz, K.; Osman, R. *Ann. N.Y. Acad. Sci.* **1994**, *726*, 71–91.
52. Miaskiewicz, K.; Miller, J.; Ornstein, R.; Osman, R. *Biopolymers* **1995**, *35*, 113–24.
53. Weiss, B.; Cunningham, R. P. *J. Bacteriol.* **1985**, *162*, 607–610.
54. Jiang, D.; Hatahet, Z.; Melamede, R. J.; Kow, Y. W.; Wallace, S. S. *J. Biol. Chem.* **1997**, *272*, 32230–32239.
55. Roldan-Arjona, T.; Anselmino, C.; Lindahl, T. *Nucleic Acids Res.* **1996**, *24*, 3307–3312.
56. Hilbert, T. P.; Boorstein, R. J.; Kung, H. C.; Bolton, P. H.; Xing, D.; Cunningham, R. P.; Teebor, G. W. *Biochemistry* **1996**, *35*, 2505–2511.
57. Sarker, A. H.; Ikeda, S.; Nakano, H.; Terato, H.; Ide, H.; Imai, K.; Akiyama, K.; Tsutsui, K.; Bo, Z.; Kubo, K.; Yamamoto, K.; Yasui, A.; Yoshida, M. C.; Seki, S. *J. Mol. Biol.* **1998**, *282*, 761–774.
58. Hilbert, T. P.; Chaung, W.; Boorstein, R. J.; Cunningham, R. P.; Teebor, G. W. *J. Biol. Chem.* **1997**, *272*, 6733–6740.
59. Aspinwall, R.; Rothwell, D. G.; Roldan-Arjona, T.; Anselmino, C.; Ward, C. J.; Cheadle, J. P.; Sampson, J. R.; Lindahl, T.; Harris, P. C.; Hickson, I. D. *Proc. Natl. Acad. Sci. U.S.A.* **1997**, *94*, 109–114.
60. Ikeda, S.; Biswas, T.; Roy, R.; Izumi, T.; Boldogh, I.; Kurosky, A.; Sarker, A. H.; Seki, S.; Mitra, S. *J. Biol. Chem.* **1998**, *273*, 21585–21593.
61. Hazra, T. K.; Izumi, T.; Boldogh, I.; Imhoff, B.; Kow, Y. W.; Jaruga, P.; Dizdaroglu, M.; Mitra, S. *Proc. Natl. Acad. Sci. U.S.A.* **2002**, *99*, 3523–3528.
62. Hazra, T. K.; Kow, Y. W.; Hatahet, Z.; Imhoff, B.; Boldogh, I.; Mokkapati, S. K.; Mitra, S.; Izumi, T. *J. Biol. Chem.* **2002**, *277*, 30417–30420.
63. Bandaru, V.; Sunkara, S.; Wallace, S. S.; Bond, J. P. *DNA Repair (Amst)* **2002**, *1*, 517–529.
64. Katafuchi, A.; Nakano, T.; Masaoka, A.; Terato, H.; Iwai, S.; Hanaoka, F.; Ide, H. *J. Biol. Chem.* **2004**, *279*, 14464–14471.
65. Miller, H.; Fernandes, A. S.; Zaika, E.; McTigue, M. M.; Torres, M. C.; Wente, M.; Iden, C. R.; Grollman, A. P. *Nucleic Acids Res.* **2004**, *32*, 338–345.
66. Jia, L.; Shafirovich, V.; Geacintov, N. E.; Broyde, S. *Biochemistry* **2007**, *46*, 5305–5314.

Chapter 3

Structural Studies of Alkylpurine DNA Glycosylases

Emily H. Rubinson, Suraj Adhikary, and Brandt F. Eichman*

Department of Biological Sciences and Center For Structural Biology, Vanderbilt University, Nashville, TN 37232
***brandt.eichman@vanderbilt.edu**

Alkylation of DNA bases produces a broad spectrum of cytotoxic and mutagenic lesions that are removed from the genome by alkylpurine DNA glycosylases. These DNA repair enzymes exist in eukaryotes, archaea, and bacteria, and have varied, well-defined specificities for particular alkylpurine nucleobases. Crystal structures of these enzymes in complex with DNA and alkylated bases have illuminated some of the chemical determinants for selection of damage amidst a vast background of normal DNA. However, only now are we beginning to understand the basis for alkylpurine specificity. Here, we review the structures of alkylpurine DNA glycosylases determined to date. Comparison of these structures in the context of functional data provides insight into the mechanisms of alkylpurine selection and excision.

Introduction

Humans are exposed to alkylating agents from various environmental sources, including industrial processes, cigarette smoke, diet, and chemotherapy. These agents, in addition to endogenous methyl donors, chemically modify the nucleobases of DNA to produce a variety of cytotoxic and mutagenic lesions that disrupt DNA replication and thus lead to heritable diseases and cancer (reviewed in (*1*)). To maintain genomic integrity amidst the constant threat of DNA alkylation, all organisms have devised multiple DNA repair strategies to eliminate the damage. Bases methylated at exocyclic substituents (e.g., *O*^{6}-methylguanine) are directly demethylated by DNA methyltransferases, whereas ring-substituted

1-methyladenine (1mA) and 3-methylcytosine (3mC) are specifically repaired through oxidative deamination by DNA dioxygenase (reviewed in (*2*)). The majority of alkylated bases, however, are eliminated from the genome by the base excision repair (BER) pathway (reviewed in (*3*, *4*)). DNA glycosylases initiate BER by locating the damaged base and catalyzing the hydrolysis of the C1′-N glycosylic bond that links the base to the phosphoribose backbone. The resulting abasic site is further processed by apurinic (AP) endonuclease, DNA polymerase and DNA ligase, acting sequentially to restore the DNA to an undamaged state.

DNA glycosylases that remove alkylation damage have been characterized from eukaryotes, archaea, and bacteria. These include mammalian alkyladenine DNA glycosylase (AAG) (*5*, *6*), yeast methyladenine DNA glycosylase (*S. cerevisiae* MAG and *S. pombe* MagI) (*7–9*), *E. coli* 3-methyladenine (3mA) DNA glycosylase I (TAG) and II (AlkA) (*10*, *11*), *Thermotoga maritima* methylpurine DNA glycosylase II (MpgII) (*12*), *Helicobacter pylori* 3mA DNA glycosylase (MagIII) (*13*), and most recently *Bacillus cereus* AlkC and AlkD (*14*). Whereas most DNA glycosylases are specific for a single modification, alkylpurine DNA glycosylases can recognize a chemically diverse set of lesions (Figure 1), including cytotoxic 3mA, 7-methylguanine (7mG), and the highly mutagenic 1,N^6-ethenoadenine (εA), which have been detected in humans and rats after exposure to various carcinogens (*15–17*). TAG and MagIII are highly specific for 3mA and 3mG (*13*, *18*), MpgII and AlkC/D are selective for positively charged lesions 3mA and 7mG (*12*, *14*), and AlkA and AAG can excise these lesions as well as other alkylated and modified bases, including εA and hypoxanthine (Hx) (*19–21*).

Alkylpurine DNA glycosylases can be classified into three distinct superfamilies based on their three-dimensional structures (Figure 2). The first is defined by the mixed α/β globular fold of AAG (*22*), which bears no structural resemblance to any other protein in the Protein Data Bank (PDB). A second, and by far the most common structural class, is the helix-hairpin-helix (HhH) superfamily of glycosylases and includes AlkA, TAG, MagIII, and MpgII (*12*, *23–25*). These enzymes contain a HhH DNA-binding motif and a common α-helical architecture also found in bacterial endonuclease III (Endo III) and MutY, archaeal MIG, and human 8-oxoguanine (OGG1) DNA glycosylases (*26–29*). *S. cerevisiae* MAG and *S. pombe* Mag1 likely adopt the HhH fold based on sequence similarity to AlkA (*7*, *8*). A third alkylpurine DNA glycosylase architecture was identified recently from the AlkC and AlkD proteins from *Bacillus cereus*. AlkD forms a C-shaped α-helical fold from repeating HEAT motifs, and AlkC is expected to adopt a similar fold (*30*, *31*).

Although structurally divergent, AAG and HhH glycosylases have evolved a conserved base-flipping mechanism for gaining access to damaged nucleobases in DNA (reviewed in (*32*)). Base flipping is common among DNA processing enzymes, and allows the protein to correctly identify and orient the substrate for catalysis. Recognition of the substrate base is believed to proceed in two stages—processive interrogation of the DNA duplex through non-specific, electrostatic interactions, followed by base flipping of the target base into the active site of the enzyme (*33*, *34*). The active sites of AAG and HhH glycosylases consist of a concave pocket lined with aromatic side chains that base-stack

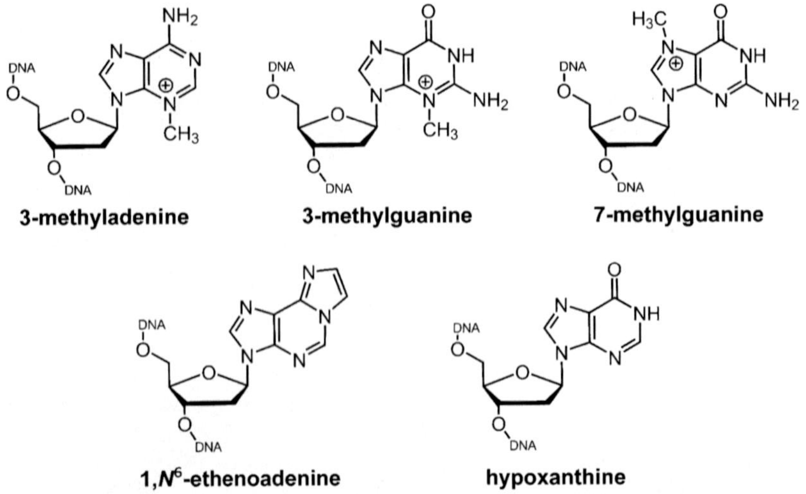

Figure 1. Nucleobases excised by alkylpurine DNA glycosylases.

with the flipped alkylpurine nucleobase, and most contain an ionizable side chain essential for catalysis. In order to stabilize the extrahelical nucleobase conformation, these glycosylases fill the gap left in the DNA by intercalating a set of side chains into the helical base stack. Although it remains to be determined if the HEAT glycosylases flip damaged bases, AlkD also contains an aromatic, electron-rich cleft but lacks identifiable intercalating residues typical of the other alkylpurine glycosylases.

Despite progress in the field, the mechanisms by which DNA glycosylases select for a particular alkyl modification are not well understood. The importance of substrate specificity is underscored by the fact that these enzymes must locate very subtle modifications among a vast excess of normal base pairs. The diversity in their structural features despite overlapping functions presents an opportunity to understand the physical and chemical determinants of DNA alkylation damage recognition and removal. In this review, we compare the alkylpurine DNA glycosylase structures determined to date, and discuss the structural implications on enzyme specificity and catalysis.

Human Alkyladenine DNA Glycosylase AAG

AAG excises a broad range of alkylpurines, including 3mA and 7mG, and has a selective preference for neutral εA and Hx (*35*). The crystal structure of AAG in complex with DNA containing an abasic pyrrolidine transition-state analog showed that AAG is a single domain protein with a mixed α/β structure and a positively charged DNA binding surface (*22*, *36*). The protein crystallized lacked the N-terminal 79 amino acids and thus the presence of a second domain is unknown. The DNA is bent at the damage site by ~22°, with B-form helical arms swung away from the protein. The pyrrolidine is rotated out of the DNA duplex and into a cavity on the protein surface. Tyr162 on the tip of a β-hairpin plugs the gap in the DNA left by the flipped pyrrolidine, and presumably stabilizes the

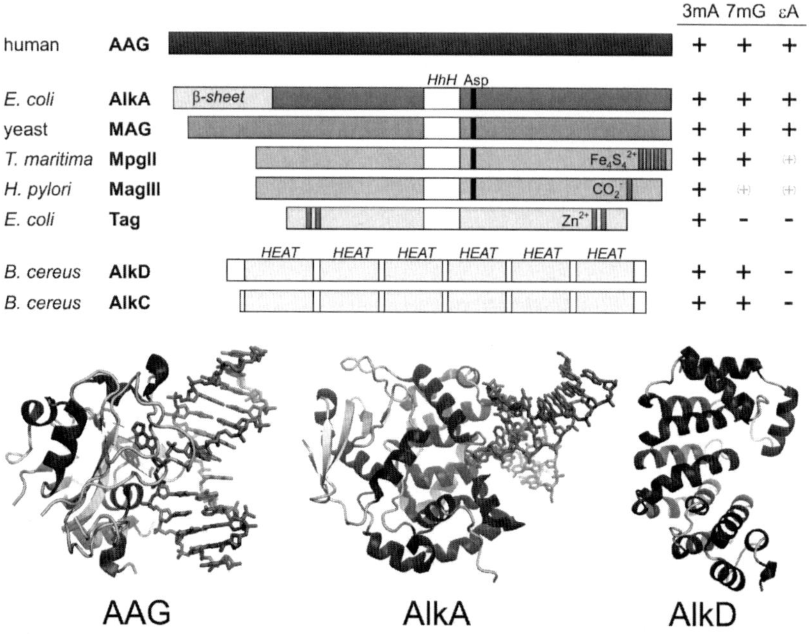

Figure 2. Structural superfamilies of alkylpurine DNA glycosylases. The three families are defined by the structural folds of human AAG, the helix-hairpin-helix (HhH) superfamily typified by AlkA, and HEAT-repeat proteins AlkC and AlkD. Substrate specificities are shown to the right of the schematic, and the crystal structures of representative proteins from each superfamily are shown at the bottom and shaded according to secondary structure. DNA is shown as sticks.

distorted conformation of the extrahelical DNA. The pyrrolidine binding pocket is lined with aromatic and polar residues. A subsequent crystal structure of AAG bound to εA-containing DNA showed the flipped εA base to be stacked between two tyrosine residues (Tyr127 and Tyr159) and His136 inside the active site cavity (Figure 3A) (*37*).

The AAG/DNA complexes have been important for directing biochemical studies aimed at understanding the molecular basis for AAG's substrate specificity and catalytic mechanism. Discrimination against normal purines is most likely due to their proper base pairing in the DNA duplex and to unfavorable interactions with exocyclic N6 and N2 amino groups inside the active site (*35*). For example, His136 donates a hydrogen bond to N6 of εA, whereas adenine cannot accept a hydrogen bond at this position. Furthermore, guanine is likely to be excluded on the basis of a steric clash between its exocyclic N2 amino group, which is absent in εA, Hx, and adenine, and the side chain of Asn169. In support of this, mutation of Asn169 gives AAG enhanced activity toward guanine (*38*). It has been suggested that AAG removes charged alkylpurine lesions because of their inherent instability and not through a structural recognition of the methyl group *per se*. Indeed, AAG's rate enhancement for excision of 3mA is one and three orders of magnitude less than that of εA and Hx, respectively (*35*). Regarding catalysis, an ordered water molecule sits adjacent to the N-glycosylic bond and is hydrogen bonded to the

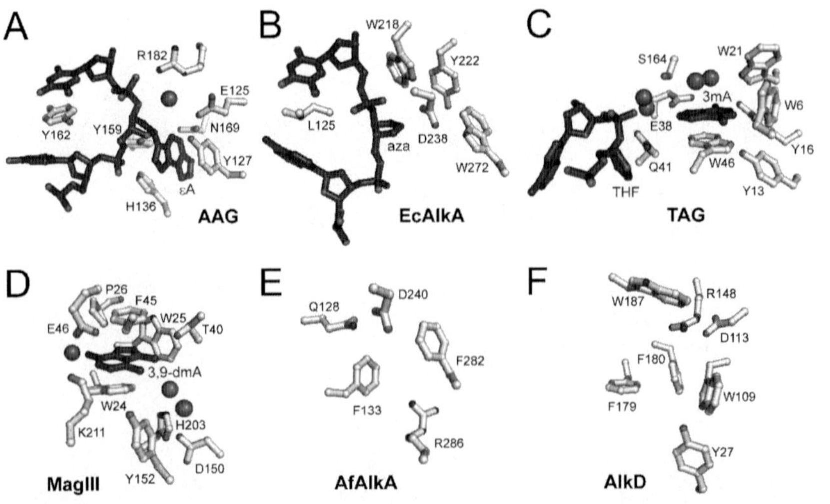

Figure 3. Active sites of alkylpurine DNA glycosylases. Protein and nucleic acid atoms are shaded black and grey, respectively. A. Human AAG in complex with εA-DNA, PDB 1ewn (37). B. E. coli AlkA bound to 1-azaribose-DNA, PDB 1diz (38). C. E. coli TAG/THF-DNA/3mA complex, PDB 2ofi (39). D. H. pylori MagIII bound to 3,9dma, PDB 1pu7 (25). E. A fulgidus AlkA, PDB 2jhj (40). F. B. cereus AlkD, PDB 3bvs (30).

side chains of Glu125 and Arg182, the carbonyl oxygen of Val262, and either the pyrrolidine N4′ or the O3′ of εA. This arrangement is consistent with Glu125 acting as a general base to deprotonate a water molecule, which may serve as a nucleophile to attack the anomeric C1′ carbon in an S_N2 catalytic mechanism (*39*).

Helix-Hairpin-Helix Superfamily

The HhH glycosylases contain two α-helical subdomains separated by an active site cleft that accommodates the flipped substrate nucleobase (Figure 4). One of these domains (helices αD-αJ) is highly conserved and contains the HhH motif (αI-αJ), a DNA binding platform utilized by hundreds of repair proteins (*40*). The HhH anchors the protein to the DNA through electrostatic interactions between main-chain atoms from the hairpin region and the phosphoribose backbone. The HhH domain also contributes a bulky group (typically a Leu, Asn, or Gln side chain) that plugs the gap in the DNA left by the flipped-out nucleotide, and a second side chain (Phe, Tyr, Leu, or Pro) that wedges between the bases opposite the flipped out nucleotide (Figure 4). Both plug and wedge residues are important for stabilizing the bent conformation of the DNA, and the wedge residue has been implicated in probing the DNA helix during the search process (*34*). The second domain, formed from the N- and C-termini (the N/C domain, helices αB-αC and αK-αM), is more varied in structure and often contains additional structural elements, including a zinc binding motif (TAG), a carbamylated lysine (MagIII), and an iron-sulfur cluster (MpgII) (Figure 4). The precise role of the N/C domain is not clear, but it is suspected that these elements help fold this domain in order to form the active site cleft.

The shape and chemical features of the active site cleft play a large role in defining the substrate specificity of these enzymes (*25*). Like AAG, HhH alkylpurine glycosylases contain aromatic, electron-rich nucleobase binding pockets that stack against ring-substituted purines. Methylated or protonated purines have enhanced π-orbital overlap between the modified base and the aromatic side chains (*42*), suggesting that the methyl group may be sensed by enhanced base stacking interactions in addition to, or in lieu of, a direct van der Waals interaction with the methyl group itself. With the exception of TAG, the HhH glycosylases contain a conserved, catalytically essential aspartic acid residue at the mouth of the active site.

E. coli AlkA

Crystal structures of *E. coli* AlkA revealed that the HhH architecture, first observed in *E. coli* Endo III (*26*), is also present in the alkylpurine DNA glycosylases (*23*, *43*). AlkA lacks the iron-sulfur cluster present in EndoIII and MutY, and instead contains an amino-terminal β-sheet domain that has no identified function but presumably stabilizes the overall fold. Crystal structures of AlkA in complex with DNA containing a 1-azaribose abasic site illuminated how DNA glycosylases utilize the HhH motif to anchor the protein to the DNA (*44*). Although the HhH does not directly participate in lesion recognition, it contributes most of the polar interactions between AlkA and the DNA. The DNA is highly distorted with a ~60° bend and widened minor groove around the site of the lesion. The 1-azaribose is rotated 180° around the phosphoribose backbone and points into a shallow cleft formed by several aromatic side chains (Figure 3B). Leu125 plugs the gap left by the flipped nucleotide in a manner similar to AAG Tyr162, and the hairpin between helices αG and αH wedge into the DNA strand opposite the lesion (Figures 3, 4, and 5).

AlkA's nucleobase binding surface is a shallow cleft that can accommodate a variety of alkylpurines. This open architecture helps explain AlkA's broad specificity. In addition, the substrate methylpurine base presumably stacks against the Trp272 indole ring, enhancing the preference of AlkA for positively charged bases. In the AlkA/DNA complex, rotation of the 1-azaribose into the active site places the N1′ nitrogen directly adjacent to the carboxylate group of the catalytic aspartate (Asp238), leaving no room for a water nucleophile necessary for an S_N2 catalytic mechanism (Figure 3B). This close proximity between the abasic site and Asp238 has led to the suggestion that AlkA utilizes an S_N1-type mechanism, whereby the ionized carboxylate stabilizes the carbocation intermediate formed on the ribose ring during nucleobase hydrolysis (*44*).

E. coli TAG

The 3mA-specific TAG enzyme is a divergent member of the HhH superfamily (*24*). Despite a conserved HhH domain, TAG lacks the conserved

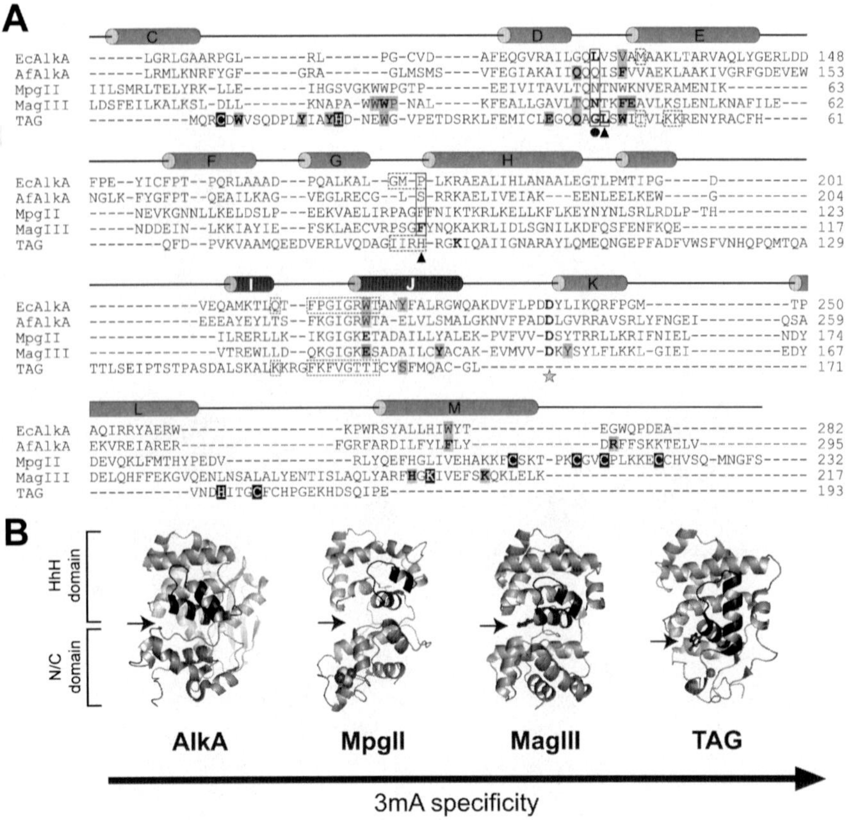

Figure 4. The helix-hairpin-helix superfamily. A. Structure based sequence alignment of E. coli AlkA, A. fulgidus AlkA, B. halodurans Mag, H. pylori MagIII, and S. typhimurium TAG is shown with secondary structure from AlkA. The MpgII sequence was aligned with MagIII. Residues that contact DNA in protein/DNA complexes of AlkA and TAG are highlighted with dashed boxes, and intercalating plug and wedge residues are boxed solid and marked beneath the sequences with a circle and triangle, respectively. Alkylpurine binding pocket residues are highlighted grey, and the catalytic aspartate is labeled with a star. Residues that coordinate ions in TAG (Zn^{2+}), MagIII (carbamylated lysine), and MpgII (iron-sulfur cluster) are shaded black. B. Crystal structures are the same as those in Figure 3 and include B. halodurans Mag, PDB 2h56 (41). The MpgII model was constructed using atomic coordinates from MagIII and MutY (1MUY, (27)) as described in the text. HhH motifs are shaded black, and the substrate base binding pockets are marked with an arrow.

catalytic aspartate present in all other HhH glycosylases, and the sequence and structure of the HhH motif itself is noticeably different (Figure 4). In addition, the N/C domain is devoid of any significant α-helical structure but rather contains a novel zinc binding motif that helps "snap" the N- and C-termini together (*45*). NMR and base perturbation studies revealed that *E. coli* TAG binds 3mA inside a deep pocket that is sterically constrained to exclude 7mG and εA bases (*46*).

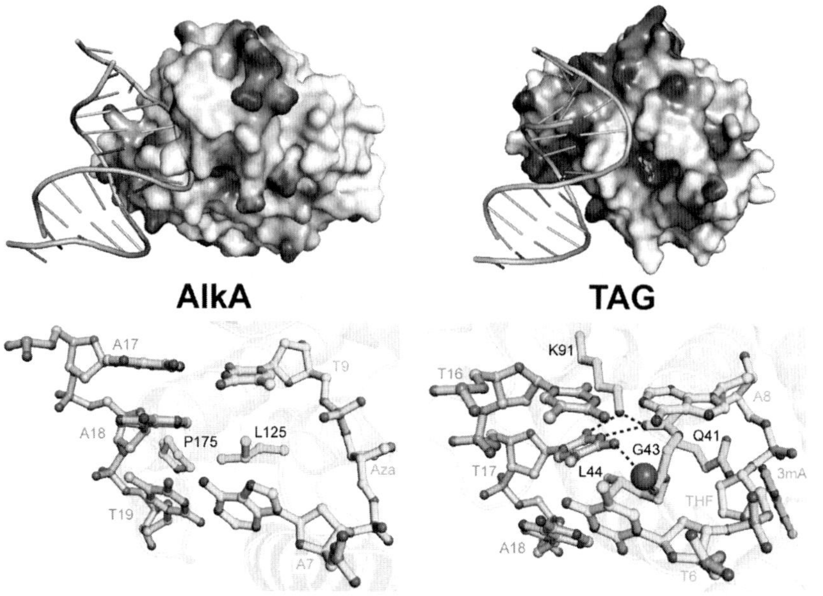

Figure 5. Comparison of TAG and AlkA DNA complexes. The overall structures of AlkA bound to 1-azaribose-containing DNA (left) and of TAG bound to THF-DNA and 3mA (right) are shown at the top, with protein rendered as an electrostatic potential surface (red, negative; blue, positive), DNA as gold cartoon, and 3mA nucleobase ball-and-stick (green carbons). At the bottom is a close-up view of the plug-and-wedge intercalation of the DNA duplexes by the proteins. A bridging water molecule in TAG is depicted as a red sphere. (see color insert)

The recent crystal structure of a TAG/DNA/3mA ternary complex provided insight into how TAG achieves is high selectivity for 3mA (*47*). *S. typhimurium* TAG, which shares 82% sequence identity (91% similarity) to the *E. coli* protein, was crystallized in the presence of free 3mA base and DNA containing a tetrahydrofuran (THF) abasic site (Figures 3C and 5). As in the AlkA/DNA complex, the HhH hairpin contributes most of the electrostatic interactions to the DNA backbone immediately 3′ to the lesion. For the first time in a DNA glycosylase structure, however, the THF moiety is not fully flipped into the active site and does not form any polar interactions with the protein. Instead, the abasic site was observed in the electron density map to interconvert between a stacked position normally found in B-DNA and one in which the ribose is partially rotated ~90° into the minor groove. The DNA is bent by ~65° as a consequence of the intercalating plug and wedge interactions, which in TAG are provided by a single hairpin loop between helices αD and αE. The main-chain of Gly43 plugs the abasic gap and the adjacent Leu44 side chain wedges between the bases across from the lesion (Figures 4 and 5). Despite the kink in the DNA, the helix remains essentially B-form as a result of the lack of specific interactions to the abasic site or to the DNA duplex on the 5′ side of the lesion. Thus, the DNA in the

TAG product complex is less distorted than in the AlkA transition state complex (Figure 5).

In the TAG/DNA crystal structure, the 3mA base resides 8 Å away from the THF moiety and deep inside the active site pocket. The 3mA ring is stacked between Trp46 and several ordered water molecules and is constrained on the sides by hydrogen bonds to Glu38 and Tyr16 and by van der Waals contacts to Trp6 (Figure 3C). Substitution of Trp46 to alanine reduced the rate of 3mA excision 10-fold with respect to wild-type TAG, further highlighting the importance of base stacking on alkylpurine glycosylase activity. The hydrogen bonds between the Glu38 carboxylate and the N6 amino and N7 imino nitrogens of 3mA suggest that 7mG is sterically excluded from the TAG active site (*46*, *47*). Interestingly, mutation of Glu38 to alanine, which should relax this constraint, did not provide TAG the ability to cleave 7mG from DNA. Rather, by two orders of magnitude with respect to the wild-type enzyme, suggesting that Glu38 is important for catalysis (*47*). A catalytic mechanism for 3mA excision has been proposed in which TAG provides a high-affinity base binding pocket that induces strain in the pre-catalytic TAG/DNA-3mA ground state complex (*46*). Release of this strain upon base hydrolysis is illustrated in the crystal structure of the TAG/DNA product complex by the large distance between 3mA and the abasic ribose and by the relatively small distortion to the conformation of the DNA helix as compared to the AlkA transition state complex (Figure 5). TAG's high specificity for 3mA, therefore, may be a result of the intrinsic instability of this lesion and the lack of a general acid or base to drive catalysis, rather than a mere steric exclusion of other bases from the active site.

S. cerevisiae MAG and *S. pombe* Mag1

S. cerevisiae MAG and *S. pombe* Mag1 share 42% and 47% overall sequence similarity to *E. coli* AlkA. Despite the similarity, MAG and Mag1 are less versatile than AlkA in their ability to excise a wide range of substrates. MAG excises 3mA, 7mG, εA, Hx, and guanine, but not oxidized substrates (e.g., *O*2-methylthymine) from DNA, while Mag1 is restricted to 3mA, 3mG, and 7mG (*19*, *48–51*). In addition, whereas MAG protects yeast cells against the toxic effects of alkylating agents and restores MMS resistance to *E. coli tag alka* mutants, *S. pombe mag1* mutants are only moderately sensitive to methylation damage (*7–9*, *52*). Interestingly, Mag1 expression is not induced by exposure to alkylating agents to the same extent as AlkA and MAG (*9*). These and other reports suggest that MAG and Mag1 play different roles in protection of yeast against alkylation damage than do the bacterial glycosylases.

The structures of MAG or Mag1 have not been determined. However, a search of MAG/Mag1 sequences against structures in the Protein Databank revealed an unpublished structure of a Mag ortholog from *Bacillus halodurans* (BhMag) determined by the Joint Center for Structural Genomics (PDB ID 2h56, (*41*)). This protein is a clear member of the HhH superfamily of DNA glycosylases (Figure 4), and shares 27% sequence identity and 65% overall similarity with the yeast proteins, although it has not been functionally characterized. Structural alignment of BhMag and AlkA shows a strong conservation in active site residues (Figure 4). The most notable difference in the base binding cleft is the presence of a methionine, which is invariant among MAG/Magl sequences, in place of the Tyr222 in AlkA (Figure 6A). It remains to be determined if this substitution accounts for the substrate specificity difference between AlkA and MAG/Magl proteins.

The asymmetric unit of the BhMag crystals contained three independent copies of the protein that adopted one of two distinct conformations. Superposition of the HhH domains of the two different conformations showed an approximate 30° rotation of the N/C-domains with respect to one another (Figure 6B). This domain displacement moves the N/C-domain into a position superimposable with AlkA, and would thus presumably allow the protein to interact more favorably with DNA. To our knowledge, this is the first observation of movement of the HhH and N/C domains in HhH glycosylases with respect to one another, and suggests that Mag orthologs might change conformation upon binding DNA. The lack of movement of the N/C-domain upon AlkA binding DNA is likely due to the presence of the unique β-sheet subdomain packed against the back-side of both HhH and N/C-domains. In the case of TAG, movement of the N/C domain was not expected since it did not contact the DNA. The effect of DNA on Mag protein structure awaits further investigation. Both MAG and Magl will be critical to our understanding of alkylation damage specificity by the highly divergent HhH family of DNA glycosylases.

H. pylori MagIII

MagIII and MpgII are two related prokaryotic alkylpurine glycosylases identified by their sequence similarity to EndoIII (*12*, *13*). The crystal structure of MagIII provided additional insight into 3mA excision and specificity (*25*). As predicted, MagIII's HhH domain structure is most similar to the iron-sulfur containing EndoIII and MutY glycosylases. Instead of a metal center, the N/C domain of MagIII contains a carbamylated lysine (Lys205) that forms extensive electrostatic interactions and likely stabilizes the protein fold (Figure 4). The interface of HhH and N/C domains forms a deep, electronegative nucleobase binding pocket lined with aromatic residues and perfectly shaped to provide a snug fit for 3mA. Structures of MagIII bound to positively charged 3,9-dimethyladenine (3,9-dmA) and neutral εA bases showed that nucleobases stack between Phe45 and Trp24 at the faces and between Trp25, Pro26, and Lys211 around the edges (Figure 3D). Other than van der Waals and π-stacking interactions, there are no specific contacts to the adenine rings like those observed in TAG. Superposition of 7mG onto the 3,9-dmA and εA rings shows that 7mG is sterically excluded from the MagIII pocket, but that the guanine ring would be accommodated by an outward rotation the Lys211 side chain. Importantly, the MagIII structures show that specific protein-nucleobase hydrogen bonds are not necessary for 3mA specificity (*25*).

In contrast to TAG, MagIII is able to excise mispaired 7mG•T and εA•C from DNA (*25*), albeit at a much lower level than AlkA and AAG (*35*, *53*). MagIII's

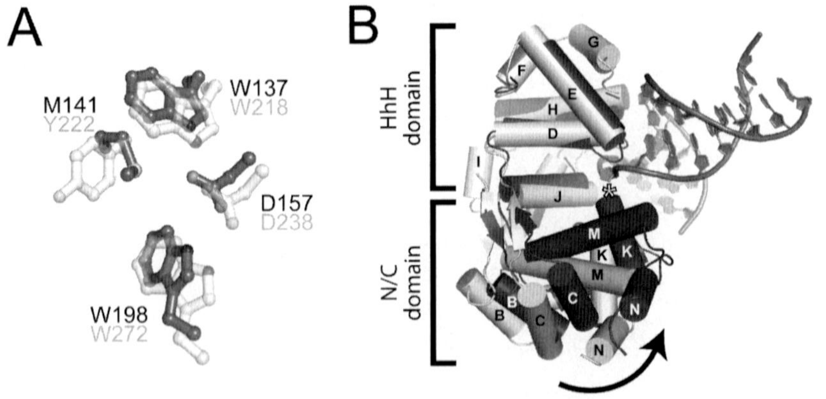

Figure 6. The crystal structure of B. halodurans Mag. A. Superposition of BhMag (black) and AlkA (grey) shows putative nucleobase binding residues are identical except for one substitution. B. Superposition of the HhH domains of the two distinct conformations of BhMag (black, grey). DNA from the AlkA/DNA crystal structure was docked onto the structure by structural alignment of the BhMag and AlkA proteins. The position of the catalytic aspartic acid is highlighted with an asterisk. The 30° rotation of the N/C-domain toward the active site of the protein places it in position to favorably interact with DNA.

weak activity for εA is likely provided by the presence of the conserved catalytic aspartate (Asp150), because mutation of this residue reduces 7mG and εA excision below measureable levels. Interestingly, the D150N mutant exhibits only a 20-fold reduction in the rate of 3mA excision. This residual 3mA activity in the aspartate mutant is further evidence that little catalytic assistance is required for hydrolysis of the labile 3mA glycosylic bond.

T. maritima MpgII

Unlike MagIII, MpgII shows robust activity toward 7mG, which is intriguing given the sequence similarity between MagIII and MpgII (Figure 4) (*12*). MpgII differs from MagIII in only two residues within the active site, Trp52 and Lys53 (Phe45 and Glu46 in MagIII, respectively). Additionally, the N/C domain of MpgII contains an iron-sulfur cluster which is absent in MagIII (Figure 4). To help understand the specificity difference between these two enzymes, we generated an MpgII model using the crystal structures of MagIII and MutY as templates (Figure 4). A chimeric MagIII-MutY template was constructed by superposition of their HhH domains followed by fusion of polypeptides from MutY residues 1-18, 171-177, 191-224 to MagIII residues 22-176 and 196-211. The MpgII sequence was threaded onto the template structure and energy minimized using the Swiss-PDBViewer and SWISS-MODEL (*54*). This model predicts that the MpgII active site is less spatially constrained than MagIII as a result of electrostatic repulsion between Lys53 and Lys204 (normally a Glu46-Lys211 salt bridge in MagIII). Substituting MagIII Glu46 with lysine to mimic the MpgII enzyme resulted in an 8-fold increase in 7mG•T activity, suggesting that steric exclusion of 7mG partially accounts MagIII's low activity toward methylguanine bases (*25*).

MpgII is the only alkylpurine-specific DNA glycosylase that contains an iron-sulfur cluster. Iron-sulfur clusters play diverse enzymatic roles and are found in a variety of DNA processing enzymes (*55–60*), although their function in DNA repair enzymes remains unclear. The iron-sulfur clusters in MutY and EndoIII glycosylases are ~15 Å from both the active site and the bound DNA (*61*, *62*). These structural observations, together with biochemical studies on MutY and EndoIII, suggest that iron-sulfur clusters play purely structural roles in DNA glycosylases (*63*, *64*). It is intriguing to speculate that the iron-sulfur cluster contributes to MpgII's enhanced activity toward more stable εA lesions.

A. fulgidus AlkA

The major single-stranded DNA alkylation products 1mA and 3mC are repaired by oxidative deamination by AlkB (*65*, *66*), a DNA dioxygenase found in bacteria and mammals but with no known orthologs in archaea. An AlkA ortholog from the archaeon *Archaeoglobus fulgidus*, AfAlkA, has been shown to excise 1mA and 3mC in addition to 3mA, 7mG, εA and Hx from DNA (*67–69*). The crystal structure of AfAlkA reveals a globular, three-domain architecture with a HhH domain similar to *E. coli* AlkA (EcAlkA) (*69*). However, the putative substrate binding pocket of AfAlkA is markedly different than that of EcAlkA (Figure 3E). The flipped substrate base is predicted to base stack between Phe133 and Phe282 in a manner similar to that observed in MagIII (Figure 3D). Substitution of either residue with alanine impairs εA and 1mA base excision, and the double mutant almost abolishes glycosylase activity, demonstrating the importance of base stacking to AfAlkA activity. Arg286 is predicted to help orient εA in the active site through hydrogen bonding, but would potentially repel the protonated amine groups of 1mA and 3mC (*69*). Mutation of Asp240, which is structurally equivalent to catalytic Asp238 in EcAlkA, results in a complete loss of function. AfAlkA represents the first reported glycosylase to have activity towards 1mA and 3mC, and further investigation will be required to determine the specificity .

HEAT Repeat Glycosylases AlkC and AlkD

Two alkylpurine DNA glycosylases, AlkC and AlkD, were identified in *Bacillus cereus* as functional complements to *E. coli* AlkA (*14*). The sequences of AlkC and AlkD are distantly related (Figure 7A) and are distinct from other known proteins. Both specifically excise positively charged bases and have no measureable activity toward εA or Hx. AlkC is highly specific for 3mA and 3mG, while AlkD also efficiently removes 7mG from DNA. The high resolution crystal structure of *Bacillus cereus* AlkD shows that the protein adopts a C-shaped globular fold composed exclusively of helical HEAT-like repeats (Figure 7B), and thus represents an unprecedented DNA glycosylase architecture (*30*). HEAT motifs are common protein binding domains that have been adapted by AlkD to

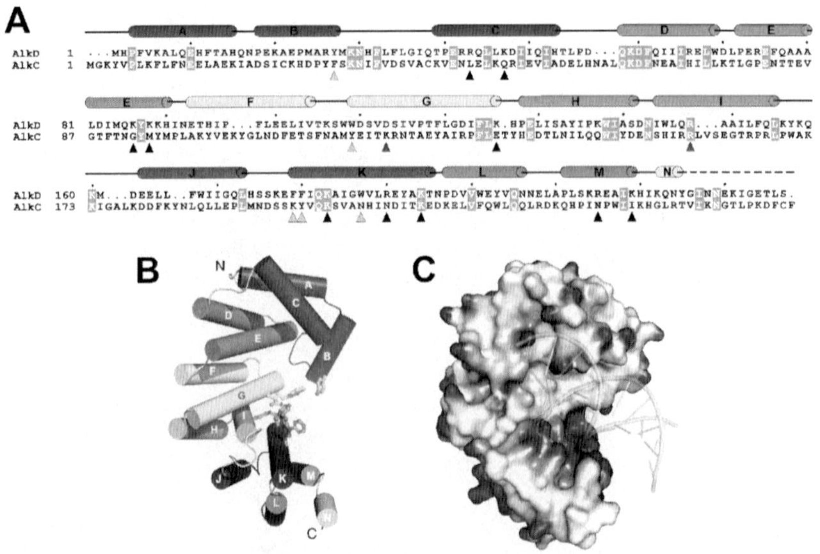

Figure 7. The AlkC/AlkD superfamily. A. Sequence alignment of B. cereus AlkC and AlkD. The secondary structure of AlkD is shown schematically at the top, and colored by heat repeat. Invariant residues are highlighted grey. Triangles denote AlkD residues identified from the crystal structure: putative active site (yellow), catalytic D113 and R148 (red), and positive charges lining the concave cleft (black). B. Crystal structure of AlkD, colored as in A. Putative active site side chains are shown as ball-and-stick. C. Electrostatic surface potential of AlkD (red, negative; blue, positive). DNA (transparent gold cartoon) modeled onto the surface highlights that the concave electropositive surface of AlkD is complementary in shape and charge to B-form DNA duplex. (see color insert)

bind DNA. The C-terminal α-helix of each HEAT repeat forms the inner, concave surface of the protein and contains lysine or arginine residues at conserved positions. Consequently, the concave surface of AlkD is positively charged and perfectly shaped to accommodate a DNA duplex (Figure 7C).

At the heart of AlkD's concave cleft is a cluster of aromatic and charged residues that resembles the active sites of other alkylpurine DNA glycosylases, implicating this region in catalysis (Figures 3F and 7B). Most notably, Asp113 and Arg148, which form an electrostatic bond at the rear of this shallow cleft, are essential for 7mG excision (*30*). Additionally, AlkD mutants with substitutions at Asp113, Arg148, Trp109, or Trp187 fail to complement the MMS sensitive phenotype of an *E. coli tag alkA* strain and decrease or abolish 3mA excision from a methylated genomic DNA substrate (*31*). Despite the evidence for an AlkD active site, the mechanism of base excision remains to be determined.

There are important structural differences which distinguish AlkD from other glycosylases and argue against a base-flipping mechanism. In all other glycosylases, the catalytic aspartate is never involved in a salt bridge and typically resides at the mouth of the active site, and is thus positioned in close proximity to the glycosylic bond of the flipped substrate base. A theoretical model of DNA docked onto AlkD illustrates that the phosphate backbone of a B-DNA duplex would electrostatically contact the Asp113 side chain. In order to accommodate a flipped base, the DNA backbone would sit farther away from the protein and consequently would lose favorable electrostatic interactions (Figure 7C). In support of this, D113N and R148A mutants cause a two-fold increase and decrease in DNA binding affinity, respectively, over wild-type AlkD. In addition, AlkD binds with the same affinity to oligonucleotides containing either G, 7mG, THF, 1-azaribose, or pyrrolidine, suggesting that the protein does not specifically interact with the lesion (*30*). These observations and the unique protein architecture imply that AlkD utilizes a novel strategy to manipulate DNA in its search for alkylpurine bases. Structures of AlkD in complex with DNA should help resolve this issue.

Summary

The structures of alkylpurine DNA glycosylases define three of the six known protein folds of DNA glycosylases, and show that various protein architectures can be used to create a DNA binding platform suitable for nucleobase excision. In addition, these structures have provided insight into enzymatic selection and hydrolysis of alkylpurines, a diverse array of lesions created from exposure of DNA to alkylating agents. Despite their diversity, the alkylpurine DNA glycosylases utilize the same general strategy for DNA damage recognition as do other glycosylases, in which amino acids near the active site are able to sense an energetic difference between modified and unmodified base pairs. Nevertheless, the alkylpurine specific active sites seem to have evolved unique mechanisms for excision of either relatively unstable, positively charged bases (e.g., cytotoxic 3mA) in addition to more stable adducts like mutagenic εA.

Acknowledgments

The authors wish to thank Patrick O'Brien for critical reading of the manuscript. Work on DNA alkylation repair in the Eichman laboratory is supported by the American Cancer Society (RSG-07-063-01-GMC), the Vanderbilt-Ingram Cancer Center (IRG-58-009-47), the Sartain-Lanier Family Foundation, and the Vanderbilt Center in Molecular Toxicology (T32 ES07028).

References

1. Friedberg, E. C.; Walker, G. C.; Siede, W.; Wood, R. D.; Schultz, R. A.; Ellenberger, T. *DNA Repair and Mutagenesis*, 2nd ed.; ASM Press: Washington, D.C., 2006.
2. Sedgwick, B. *Nat. Rev.* **2004**, *5*, 148–157.
3. Fromme, J. C.; Verdine, G. L. *Adv. Protein Chem.* **2004**, *69*, 1–41.
4. Huffman, J. L.; Sundheim, O.; Tainer, J. A. *Mutat. Res.* **2005**, *577*, 55–76.
5. Karran, P.; Lindahl, T. *Biochemistry* **1980**, *19*, 6005–6011.
6. Brent, T. P. *Biochemistry* **1979**, *18*, 911–916.

7. Chen, J.; Derfler, B.; Samson, L. *EMBO J.* **1990**, *9*, 4569–4575.
8. Berdal, K. G.; Bjoras, M.; Bjelland, S.; Seeberg, E. *EMBO J.* **1990**, *9*, 4563–4568.
9. Memisoglu, A.; Samson, L. *Gene* **1996**, *177*, 229–235.
10. Riazuddin, S.; Lindahl, T. *Biochemistry* **1978**, *17*, 2110–2118.
11. Thomas, L.; Yang, C. H.; Goldthwait, D. A. *Biochemistry* **1982**, *21*, 1162–1169.
12. Begley, T. J.; Haas, B. J.; Noel, J.; Shekhtman, A.; Williams, W. A.; Cunningham, R. P. *Curr. Biol.* **1999**, *9*, 653–656.
13. O'Rourke, E. J.; Chevalier, C.; Boiteux, S.; Labigne, A.; Ielpi, L.; Radicella, J. P. *J. Biol. Chem.* **2000**, *275*, 20077–20083.
14. Alseth, I.; Rognes, T.; Lindback, T.; Solberg, I.; Robertsen, K.; Kristiansen, K. I.; Mainieri, D.; Lillehagen, L.; Kolsto, A. B.; Bjoras, M. *Mol. Microbiol.* **2006**, *59*, 1602–1609.
15. Holt, S.; Yen, T. Y.; Sangaiah, R.; Swenberg, J. A. *Carcinogenesis* **1998**, *19*, 1763–1769.
16. Shuker, D. E.; Bailey, E.; Parry, A.; Lamb, J.; Farmer, P. B. *Carcinogenesis* **1987**, *8*, 959–962.
17. Shuker, D. E.; Farmer, P. B. *Chem. Res. Toxicol.* **1992**, *5*, 450–460.
18. Bjelland, S.; Bjoras, M.; Seeberg, E. *Nucleic Acids Res.* **1993**, *21*, 2045–2049.
19. Saparbaev, M.; Kleibl, K.; Laval, J. *Nucleic Acids Res.* **1995**, *23*, 3750–3755.
20. McCarthy, T. V.; Karran, P.; Lindahl, T. *EMBO J.* **1984**, *3*, 545–550.
21. Bjelland, S.; Birkeland, N. K.; Benneche, T.; Volden, G.; Seeberg, E. *J. Biol. Chem.* **1994**, *269*, 30489–30495.
22. Lau, A. Y.; Schärer, O. D.; Samson, L.; Verdine, G. L.; Ellenberger, T. *Cell* **1998**, *95*, 249–258.
23. Labahn, J.; Schärer, O. D.; Long, A.; Ezaz-Nikpay, K.; Verdine, G. L.; Ellenberger, T. E. *Cell* **1996**, *86*, 321–329.
24. Drohat, A. C.; Kwon, K.; Krosky, D. J.; Stivers, J. T. *Nat. Struct. Biol.* **2002**, *9*, 659–664.
25. Eichman, B. F.; O'Rourke, E. J.; Radicella, J. P.; Ellenberger, T. *EMBO J.* **2003**, *22*, 4898–4909.
26. Kuo, C. F.; McRee, D. E.; Fisher, C. L.; O'Handley, S. F.; Cunningham, R. P.; Tainer, J. A. *Science* **1992**, *258*, 434–440.
27. Guan, Y.; Manuel, R. C.; Arvai, A. S.; Parikh, S. S.; Mol, C. D.; Miller, J. H.; Lloyd, S.; Tainer, J. A. *Nat. Struct. Biol.* **1998**, *5*, 1058–1064.
28. Bruner, S. D.; Norman, D. P.; Verdine, G. L. *Nature* **2000**, *403*, 859–866.
29. Mol, C. D.; Arvai, A. S.; Begley, T. J.; Cunningham, R. P.; Tainer, J. A. *J. Mol. Biol.* **2002**, *315*, 373–384.
30. Rubinson, E. H.; Metz, A. H.; O'Quin, J.; Eichman, B. F. *J. Mol. Biol.* **2008**, *381*, 13–23.
31. Dalhus, B.; Helle, I. H.; Backe, P. H.; Alseth, I.; Rognes, T.; Bjoras, M.; Laerdahl, J. K. *Nucleic Acids Res.* **2007**, *35*, 2451–2459.
32. Roberts, R. J.; Cheng, X. *Annu. Rev. Biochem.* **1998**, *67*, 181–198.
33. Stivers, J. T.; Jiang, Y. L. *Chem. Rev.* **2003**, *103*, 2729–2759.
34. Banerjee, A.; Santos, W. L.; Verdine, G. L. *Science* **2006**, *311*, 1153–1157.

35. O'Brien, P. J.; Ellenberger, T. *J. Biol. Chem.* **2004**, *279*, 9750–9757.
36. Scharer, O. D.; Nash, H. M.; Jiricny, J.; Laval, J.; Verdine, G. L. *J. Biol. Chem.* **1998**, *273*, 8592–8597.
37. Lau, A. Y.; Wyatt, M. D.; Glassner, B. J.; Samson, L. D.; Ellenberger, T. *Proc. Natl. Acad. Sci. U.S.A.* **2000**, *97*, 13573–13578.
38. Connor, E. E.; Wyatt, M. D. *Chem. Biol.* **2002**, *9*, 1033–1041.
39. O'Brien, P. J.; Ellenberger, T. *Biochemistry* **2003**, *42*, 12418–12429.
40. Doherty, A. J.; Serpell, L. C.; Ponting, C. P. *Nucleic Acids Res.* **1996**, *24*, 2488–2497.
41. Protein Data Bank, Joint Center for Structural Genomics, PDB ID 2H56, 2006.
42. Ishida, T.; Shibata, M.; Fujii, K.; Inoue, M. *Biochemistry* **1983**, *22*, 3571–3581.
43. Yamagata, Y.; Kato, M.; Odawara, K.; Tokuno, Y.; Nakashima, Y.; Matsushima, N.; Yasumura, K.; Tomita, K.; Ihara, K.; Fujii, Y.; Nakabeppu, Y.; Sekiguchi, M.; Fujii, S. *Cell* **1996**, *86*, 311–319.
44. Hollis, T.; Ichikawa, Y.; Ellenberger, T. *EMBO J.* **2000**, *19*, 758–766.
45. Kwon, K.; Cao, C.; Stivers, J. T. *J. Biol. Chem.* **2003**, *278*, 19442–19446.
46. Cao, C.; Kwon, K.; Jiang, Y. L.; Drohat, A. C.; Stivers, J. T. *J. Biol. Chem.* **2003**, *278*, 48012–48020.
47. Metz, A. H.; Hollis, T.; Eichman, B. F. *EMBO J.* **2007**, *26*, 2411–2420.
48. Saparbaev, M.; Laval, J. *Proc. Natl. Acad. Sci. U.S.A.* **1994**, *91*, 5873–5877.
49. Bjoras, M.; Klungland, A.; Johansen, R. F.; Seeberg, E. *Biochemistry* **1995**, *34*, 4577–4582.
50. Berdal, K. G.; Johansen, R. F.; Seeberg, E. *EMBO J.* **1998**, *17*, 363–367.
51. Alseth, I.; Osman, F.; Korvald, H.; Tsaneva, I.; Whitby, M. C.; Seeberg, E.; Bjoras, M. *Nucleic Acids Res.* **2005**, *33*, 1123–1131.
52. Memisoglu, A.; Samson, L. *J. Bacteriol.* **2000**, *182*, 2104–2112.
53. O'Brien, P. J.; Ellenberger, T. *J. Biol. Chem.* **2004**, *279*, 26876–26884.
54. Guex, N.; Peitsch, M. C. *Electrophoresis* **1997**, *18*, 2714–2723.
55. Flint, D. H.; Allen, R. M. *Chem. Rev.* **1996**, *96*, 2315–2334.
56. Cunningham, R. P.; Asahara, H.; Bank, J. F.; Scholes, C. P.; Salerno, J. C.; Surerus, K.; Munck, E.; McCracken, J.; Peisach, J.; Emptage, M. H. *Biochemistry* **1989**, *28*, 4450–4455.
57. Michaels, M. L.; Pham, L.; Nghiem, Y.; Cruz, C.; Miller, J. H. *Nucleic Acids Res.* **1990**, *18*, 3841–3845.
58. Rudolf, J.; Makrantoni, V.; Ingledew, W. J.; Stark, M. J.; White, M. F. *Mol. Cell* **2006**, *23*, 801–808.
59. Weiner, B. E.; Huang, H.; Dattilo, B. M.; Nilges, M. J.; Fanning, E.; Chazin, W. J. *J. Biol. Chem.* **2007**, *282*, 33444–33451.
60. Klinge, S.; Hirst, J.; Maman, J. D.; Krude, T.; Pellegrini, L. *Nat. Struct. Mol. Biol.* **2007**, *14*, 875–877.
61. Fromme, J. C.; Verdine, G. L. *EMBO J.* **2003**, *22*, 3461–3471.
62. Fromme, J. C.; Banerjee, A.; Huang, S. J.; Verdine, G. L. *Nature* **2004**, *427*, 652–656.
63. Fu, W.; O'Handley, S.; Cunningham, R. P.; Johnson, M. K. *J. Biol. Chem.* **1992**, *267*, 16135–16137.

64. Golinelli, M. P.; Chmiel, N. H.; David, S. S. *Biochemistry* **1999**, *38*, 6997–7007.
65. Falnes, P. O.; Johansen, R. F.; Seeberg, E. *Nature* **2002**, *419*, 178–182.
66. Trewick, S. C.; Henshaw, T. F.; Hausinger, R. P.; Lindahl, T.; Sedgwick, B. *Nature* **2002**, *419*, 174–178.
67. Birkeland, N. K.; Anensen, H.; Knaevelsrud, I.; Kristoffersen, W.; Bjoras, M.; Robb, F. T.; Klungland, A.; Bjelland, S. *Biochemistry* **2002**, *41*, 12697–12705.
68. Mansfield, C.; Kerins, S. M.; McCarthy, T. V. *FEBS Lett.* **2003**, *540*, 171–175.
69. Leiros, I.; Nabong, M. P.; Grosvik, K.; Ringvoll, J.; Haugland, G. T.; Uldal, L.; Reite, K.; Olsbu, I. K.; Knaevelsrud, I.; Moe, E.; Andersen, O. A.; Birkeland, N. K.; Ruoff, P.; Klungland, A.; Bjelland, S. *EMBO J.* **2007**, *26*, 2206–2217.

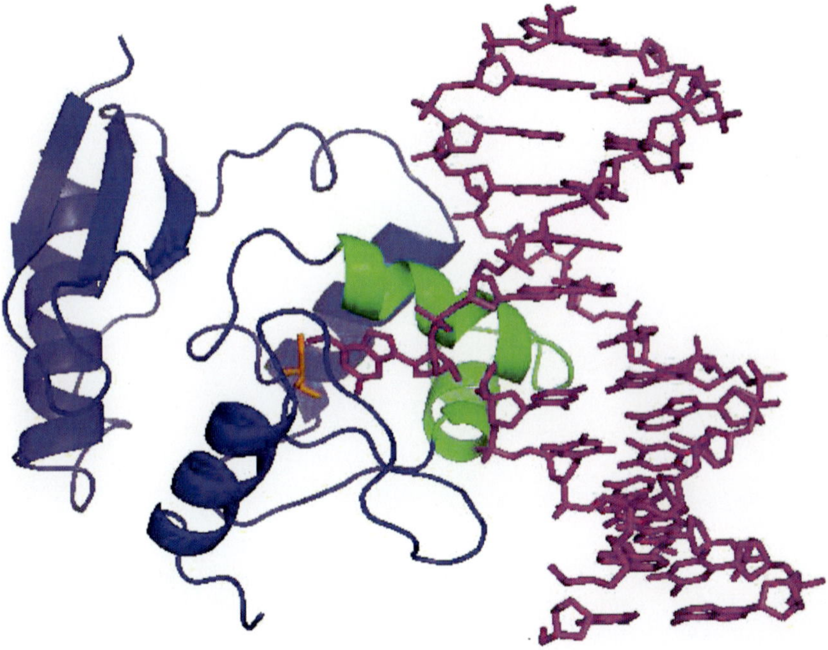

Figure 3. AGT (blue) bound to O^6-Me-dG containing DNA (pink). AGT binds DNA through a helix-turn-helix (HTH) motif (green). The adducted nucleotide is flipped into the active site of the protein where Cys145 (orange) is positioned for S_N2-type removal of the O^6-alkyl group of guanine. (Protein Data Bank accession code 1T38; (26))

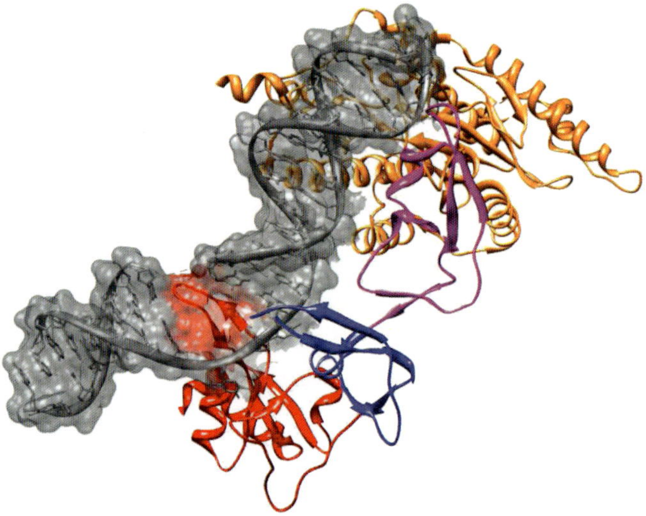

Figure 2. Crystal structure of Rad4 bound to damaged DNA. The Rad4 DNA binding fragment was co-crystallized bound to a 24 base pair duplex substrate containing a cyclobutane pyrimidine dimer (CPD) lesion located in a 2-nucleotide mismatch (PDB ID 2QSF). The TGD (gold) and BHD1 (pink) domains form a double-stranded DNA binding clamp engaging the substrate 3' with respect the to lesion. The BHD2 (blue) and BHD3 (red) domains bind the single-stranded DNA region resulting from thermodynamic destabilization of the helix by the presence of the CPD lesion. The β-hairpin loop of BHD3 inserts between the strands rotating two base pairs out of the duplex. Rad4 does not make any contact with the damaged nucleotides, which are not observed in the crystal structure and are believed to be dynamically disordered.

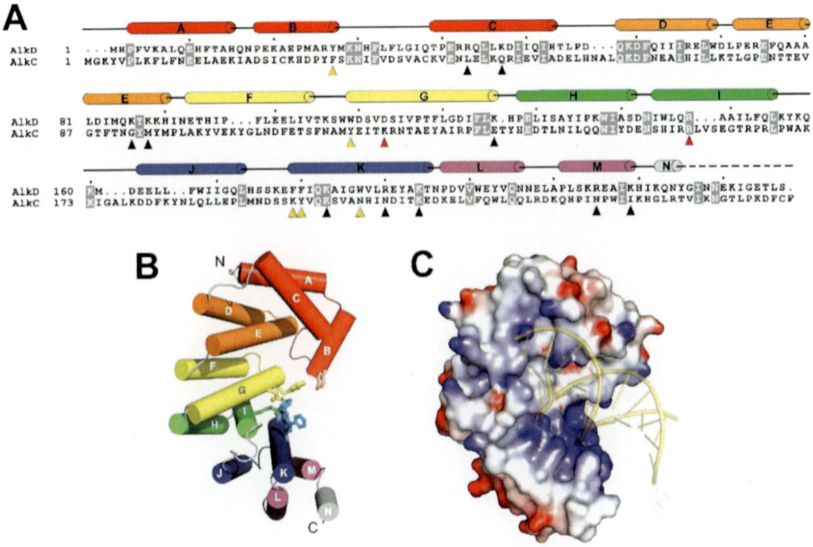

Figure 7. The AlkC/AlkD superfamily. A. Sequence alignment of B. cereus AlkC and AlkD. The secondary structure of AlkD is shown schematically at the top, and colored by heat repeat. Invariant residues are highlighted grey. Triangles denote AlkD residues identified from the crystal structure: putative active site (yellow), catalytic D113 and R148 (red), and positive charges lining the concave cleft (black). B. Crystal structure of AlkD, colored as in A. Putative active site side chains are shown as ball-and-stick. C. Electrostatic surface potential of AlkD (red, negative; blue, positive). DNA (transparent gold cartoon) modeled onto the surface highlights that the concave electropositive surface of AlkD is complementary in shape and charge to B-form DNA duplex.

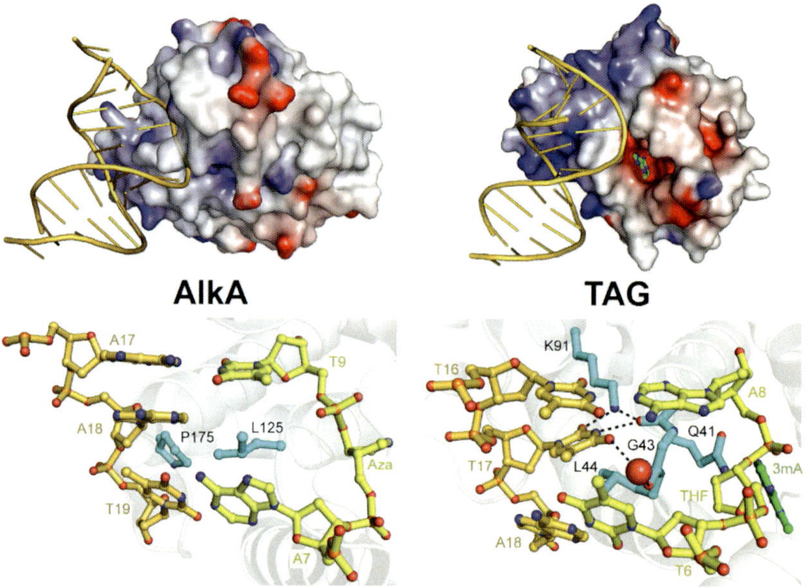

Figure 5. Comparison of TAG and AlkA DNA complexes. The overall structures of AlkA bound to 1-azaribose-containing DNA (left) and of TAG bound to THF-DNA and 3mA (right) are shown at the top, with protein rendered as an electrostatic potential surface (red, negative; blue, positive), DNA as gold cartoon, and 3mA nucleobase ball-and-stick (green carbons). At the bottom is a close-up view of the plug-and-wedge intercalation of the DNA duplexes by the proteins. A bridging water molecule in TAG is depicted as a red sphere.

Chapter 4

Dynamics in Uracil Base Excision Repair

Intrinsic Dynamics Drive the Search for a Damaged Base

Joshua I. Friedman and James T. Stivers*

Department of Pharmacology and Molecular Sciences, Johns Hopkins University School of Medicine, Baltimore, MD 21205
***jstivers@jhmi.edu**

The base excision repair enzyme human uracil DNA glycosylase (hUNG) exhibits dynamic fluctuations when bound to DNA that are consistent with transitioning between linear scanning and a pausing mode in which extrahelical thymine and uracil bases are interrogated. The absence of these motions in free UNG suggests that the enzyme uses the favorable free energy of DNA binding to loosen its own structure and activate dynamic modes necessary for the identification of damaged uracil bases.

Introduction

The first theory of enzyme function was proposed by Emil Fischer who hypothesized in the late nineteenth century that enzymes and substrates were complimentary structures that fit together like a 'lock and key' in order to carry out catalysis (*1*). Linus Pauling later refined this idea by suggesting that enzymes are complementary in structure to the transition states of their reactions and thereby lower the free energy barrier to catalysis (*2*). The development of high resolution diffraction techniques for molecular structure determination reinforced a perception of enzymes as rigid structural scaffolds via the proliferation of singular representations of protein structure.

However, crystallographic models of enzymes in multiple conformations as well as numerous biochemical inferences have made it clear that proteins are capable of undergoing biologically relevant motions (*3–5*). With relatively recent advances in NMR spectroscopy, computer modeling, and single molecule techniques, the direct investigation of protein dynamics have become possible (*6–8*). These relatively new methods have painted a quite different picture of enzyme function, where dynamic fluctuations in structure, occurring over many timescales, are envisioned as intrinsic components of enzyme structure, regulation, and catalytic function.

In this review, we describe recent NMR studies that have uncovered the role DNA and enzyme dynamics play in the recognition and excision of a uracil base from duplex DNA by the enzyme uracil DNA glycosylase (*9*, *10*). Using such NMR spectroscopic approaches, we have obtained the first glimpses into the inner workings of this key human DNA base excision repair enzyme as it patrols DNA looking for sites of DNA damage.

The Search Problem

The problem of identifying a single damaged base in a vast genome of undamaged, yet nearly isomorphic bases, represents a major obstacle to efficient DNA repair. Repair enzymes must be promiscuous enough to interrogate the entire genome for damage, but must not be so indiscriminate that they bind tightly to, or worse yet act upon, undamaged DNA. As the enzymatic process of damage search and recognition is accomplished without the direct input of cellular metabolic energy, the search must be driven entirely by Brownian motion (*11*). In this regard, two distinct thermally driven intramolecular search mechanisms may be envisioned: one dimensional sliding and three dimensional hopping along the DNA chain (*11*, *12*). We first consider the merits of these two mechanisms with respect to the efficiency of locating rare sites of DNA damage. It should be apparent after this discussion that the experimentally observed search mechanism of UNG is exquisitely well-suited for detecting damaged sites.

One-Dimensional DNA Sliding

It is well known that stochastic linear diffusion along a DNA chain has significant limitations for locating diffuse sites of damage within genome size DNA (*13*, *14*). This conclusion is based on the physics of linear diffusion, where thermal fluctuations are equally likely to push an enzyme forward along the DNA backbone towards a site of damage, as they are to push it backward away from the damage site. Accordingly, an enzyme that diffuses solely by a sliding mechanism must take n^2 steps to move n base pairs along the DNA chain (*13*, *15*, *16*). This property of one-dimensional linear scanning means that a sliding enzyme will redundantly search the same short stretch of a DNA chain, while rarely sampling distant parts of the genome that may also contain damage sites.

In theory, a search mechanism involving DNA chain sliding could be used to investigate long stretches of the DNA chain if the sliding rate were very fast. Indeed the theoretical maximum one dimensional sliding rate of an enzyme is quite rapid (~ 10^7 bp s^{-1}) (*17*), but rapid sliding is problematic for a number of practical reasons. For one, a rapidly sliding enzyme cannot interact too tightly with the nonspecific DNA sliding beneath it, otherwise its sliding movement would be hindered. This requirement must be balanced with the competing requirement that the enzyme not bind so loosely that it is on average not bound to

DNA at all. Thus, an optimized DNA repair enzyme would need to balance the weak interactions required for efficient linear diffusion with the requirement that it remain in contact with DNA. Even if a satisfactory compromise between these requirements could be found, a DNA repair enzyme must do more than simply slide over sites of damage--it must also occasionally pause along the DNA chain to allow time for the identification of a damaged base. The process of pausing would seem to require that the enzyme transiently alter its conformation such that stronger interactions with the duplex are formed. For these interactions to be efficient at halting the enzyme, they must be formed on the same time scale competitive with the sliding rate, and must also possess features that facilitate damage recognition. If these criteria are not met, then many base pairs would be passed over without the opportunity for damage detection. This network of competing and contradictory requirements makes pure one-dimensional sliding an unattractive mechanism to search DNA.

Three-Dimensional DNA Hopping

Given the limitations of linear diffusion, efficient surveillance of sites distant in sequence space would be facilitated if an enzyme could occasionally completely dissociate from the DNA and then proceed through three-dimensional space to another site on the DNA chain, a process known as 'hopping'. Like DNA scanning, hopping is stochastic process, meaning that a hopping enzyme is just as likely to move away from a given location as it is to get closer, resulting in relatively small mean displacements per unit time.

However, in the crowded environment of cellular nuclei where DNA strands are folded back on themselves, even a small displacement in three-dimensional space could lead to a very large displacement in sequence space. DNA hopping is thus well-suited to allow enzymes to sample distant parts of the genome, but by itself, fails to support efficient surveillance. This conclusion derives from the vanishingly small likelihood that a hopping enzyme would land directly in register with an extremely rare damaged base.

DNA Sliding and Hopping

The above considerations lead one to embrace a mixed model for site recognition that incorporates the best features of local linear scanning and long range hopping. The utilization of both sliding and hopping pathways has a synergistic effect on the ability of repair enzymes to effectively patrol the genome. DNA hopping enables enzymes to act over large swaths of primary sequence and avoid DNA bound protein obstacles, while DNA sliding enables the enzyme to rigorously interrogate short stretches of DNA surrounding each landing site. The specific rates k_{off} and k_{slide} have likely evolved in tandem, and in part, define a contour surface which describes the 'efficiency' of the search process (Figure 1). The specific shape of the contour surface describes the efficiency of repair (vertical axis) for a given set of sliding (k_{slide}) and hopping rates (k_{off}). In the case of uracil base excision repair, the recognition step occurs during the act of flipping the uracil base from the duplex. Thus, the rate of uracil flipping places physical constraints on how long UNG needs to be associated with a site, or pause, in order to determine if a damaged base has been encountered. The remainder of this review summarizes the body of experimental evidence that supports the above mechanistic view for uracil damage recognition.

DNA Base Pair Dynamics

The familiar structure of DNA double helix arises from the cumulative effects of base stacking, hydrogen bonding, ionic, and solvent interactions. These interactions are relatively weak on a per residue basis, and consequently, the local structure of DNA is easily perturbed at ambient temperatures. This plasticity in structure is as important a structural characteristic of DNA as the average structure depicted in crystallographic models, and must be considered in any model describing DNA damage recognition (*9*, *19*).

The DNA structural transition most relevant to base excision repair is that of base pair opening or 'breathing', as this motion is along the reaction coordinate for base flipping by DNA glycosylase enzymes (Figure 2). This breathing event is characterized by the interruption of interstrand Watson-Crick hydrogen bonds and rotation of the nucleoside about the phosphate backbone. Base pair breathing can be directly probed using nuclear magnetic resonance (NMR) magnetization transfer experiments that measure the chemical exchange of water protons with the imino protons of T or G bases. The principle of the experiment is that imino protons are sequestered in the duplex by interstrand hydrogen bonds, and that exposure of the imino proton upon base pair opening allows chemical exchange with water protons to occur (Figure 2). The NMR experiment involves magnetically labeling bulk water protons, and then following their rate of exchange into the imino sites from which the base pair opening rate can be extracted provided that the exchange step is sufficiently rapid (*20*). Such experiments have established that T/A base pairs in DNA open at a rapid rate approximated at about 8,000 s^{-1} under physiological conditions (*14*, *21*) and these extrahelical states exist for approximately ten microseconds before reclosing (*21*, *22*). The spontaneous opening rates are faster than any subsequent step in the enzymatic reaction, and thus provide kinetically competent motions to initiate damage recognition.

NMR imino proton exchange measurements have revealed a great deal about the role of intrinsic DNA dynamics in the uracil search process by UNG (*23–25*). A key finding was that UNG did not accelerate the opening rate of normal DNA base pairs as compared to the opening rates in the absence of the enzyme. This result provided strong support for a mechanism in which the intrinsic opening dynamics of normal and damaged DNA bases provided the initial seeding motions to initiate extrahelical base binding. Although UNG did not alter the base pair opening rates, it did increase the equilibrium population of extrahelical T and U bases by slowing the closure rate by about 25-fold (*23*, *24*) (Figure 2). These studies also revealed that UNG had no effect on the opening or closing rates of isolated G/C base pairs. A mechanism has thus emerged from these NMR measurements where UNG inspects thymine bases in an extrahelical

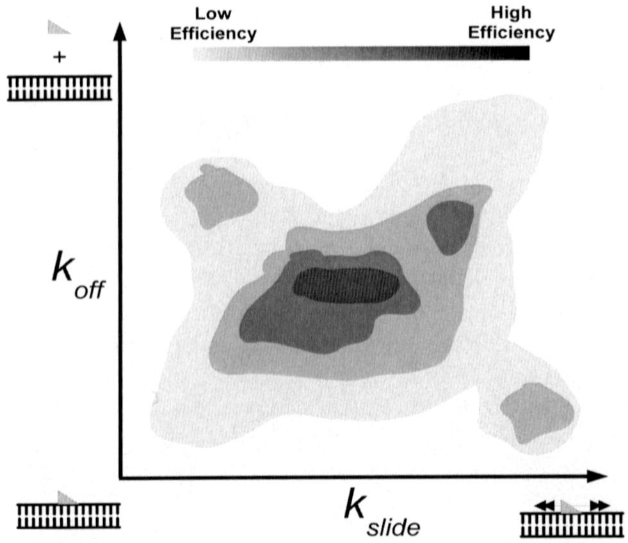

Figure 1. Contour plot representation of a hypothetical 'efficiency' landscape for an enzyme patrolling the genome for damage. The plot shows how the optimal efficiency will be realized when the off-rate and sliding rates are matched to the forward rate of damage recognition. The absolute values for these parameters will depend on the copy number of the enzyme, its catalytic power, and the size of the genome being surveyed (12, 18).

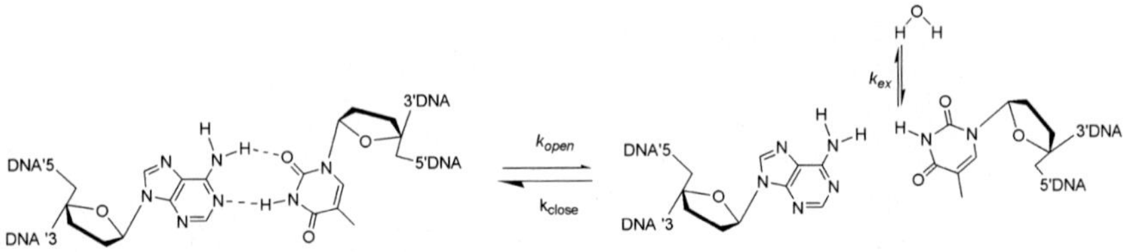

Figure 2. Opening of a T/A base pair resulting in exchange of the imino proton of the thymine base.

binding site during its search for uracil. Such a site was later established by X-ray crystallography, where a transiently extrahelical thymine was trapped in a surface pocket on UNG (an 'exosite') that was sterically matched to the size of pyrimidine bases, and possessed hydrogen bond donor and acceptor groups that were complementary to the Watson-Crick edge of thymine and uracil bases (*25*). Since thymine has a bulky 5-methyl group that prevents it from entering the active site pocket (*26*), the data indicated that this extrahelical stabilizing 'exosite' allowed both binding and discrimination between thymine and uracil.

The DNA dynamic studies and the structural model of the UNG 'exosite' complex are compatible with a mechanism for initial recognition of uracil that involves transient base pair breathing and enzymatic trapping of the nascent extrahelical state. This mechanism, although supported by many experimental measurements, dictated several mechanistic requirements. First, the short lifetime of the extrahelical base required that UNG already be present at the site and in the correct register for exosite binding to occur (diffusion from solution is not a kinetically competent mechanism) (*19*). Thus, pre-association of UNG with DNA near the site was required, followed by rapid scanning along the DNA chain to the site, a strategy identical to that of the mixed model of DNA sliding/hopping outlined above. Second, it was difficult to imagine that UNG slid on DNA using the binding mode observed in the exosite structure. This difficulty arises because the enzyme is found in a closed conformation, with the side chain alkyl group of a leucine residue intercalated into the minor groove opposite to the extrahelical thymine (*25*, *27*). These considerations led to the hypothesis that UNG must have dynamic properties that allow it to sample an open state competent for rapid translocation on the DNA, and a closed state capable of pausing along the DNA strand and stabilizing the extrahelical base. For maximum efficiency, this dynamic sampling of conformational states should happen on the timescales of the translocation rate and the binding lifetime of UNG on undamaged DNA. Recent measurements of the translocation mechanism of *E. coli* UNG have provided reasonable values for the average scanning distance (~ 10 bp), and its average bound lifetime per association event (~ 5 ms at 37 °C). These estimates provided the timescale for the anticipated dynamic motions of UNG, and this timescale was appropriate for NMR dynamic studies.

NMR and Enzyme Search Dynamics

Molecular motions affect the individual magnetic nuclei in an NMR sample slightly differently, and overtime, these motions can interfere with the ability of chemically identical nuclei to collectively generate the observed NMR signal (*28*). These motions generate time-varying dipolar fields that serve to increase the decay rate (R_2) of the observed NMR signal (the free induction decay or FID). The rate of decay of the FID is manifest as the line width (R_2/π) of the signal in the NMR spectrum. By appropriate design of an NMR experiment, one can learn a great deal about the dynamics of individual magnetic nuclei in proteins (^{15}N, ^{13}C, ^{1}H) by measuring the effect of these motions on the decay rate of the FID. The additional line broadening effect of nuclear dynamics arises from the magnetic nucleus rapidly sampling different chemical environments each with a different frequency (ω). This additional line broadening is known as the "exchange" contribution to the linewidth, which also makes the FID decay more rapidly than in the absence of exchange. This basic phenomenon provides the basis for some very useful NMR dynamic experiments that probe motions on the millisecond time scale (*29*, *30*).

Dynamics in Free UNG

The backbone NMR assignments of free human UNG were obtained by standard triple resonance NMR methods (*31*), and provided the atom specific assignments needed for backbone ^{15}N dynamic measurements (*32*). The salient observation for free UNG was that the enzyme was dynamically inert over

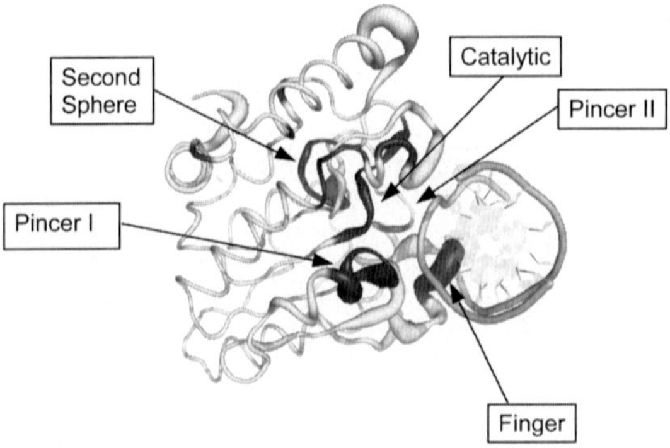

Figure 3. UNG is shown bound to non-target DNA (pdb 2OXM) (25) looking down the double helical axis of the DNA fragment. Five regions of the enzyme were identified as being dynamically active when complex with DNA and are labeled and colored black above. The width of the enzyme structure is proportional to the chemical shift perturbation of the enzyme induced by binding DNA. The DNA is held in place via specific hydrogen bonds between serine residues in Pincer regions I & II of UNG to the phosphate backbone of DNA at loci immediately surrounding the nucleobase under investigation. A structure termed finger region intercalates into the DNA base stack of dynamically opening bases slows the rate of base closure via steric exclusion. The catalytic strand, which runs deep within the protein core, contains the necessary chemical moieties for glycolytic cleavage of fully flipped uracil. A second sphere strand also shows dynamics and runs anti-parallel to the catalytic stand and threads deeper within the protein hydrophobic core.

a wide range of relevant timescales (ms-ns). Accordingly, the ^{15}N R_2 values using the CPMG relaxation method were barren of exchange contributions to the linewidths, with only a few residues exhibiting any measurable dynamics. The few notable exceptions were Ser169, Ser247 and Tyr248, which exhibited evidence of millisecond timescale dynamics in the free enzyme (*25*). Collectively these residues form opposing protrusions which hydrogen bond to the DNA strand 3′ and 5′ of the extrahelical thymine base in the exosite of UNG (*27*).

Dynamics of the Damage Search Complex

Addition of an undamaged ten base-pair oligonucleotide to the NMR sample of UNG caused perturbations in the backbone amide chemical shifts of a small group of residues (*31*). The affected amino acids mapped exclusively to the DNA binding site depicted in crystallographic structures, and thus justify the direct comparison of the NMR results with the x-ray structural model of UNG interacting with non-target DNA. Within the DNA complex, a distinct group of residues displayed exaggerated linewidths indicative of chemical exchange between multiple environments and were attributed to conformational rearrangements within the UNG-DNA complex that likely reflect the dynamic properties of the enzyme during the damage search and recognition process.

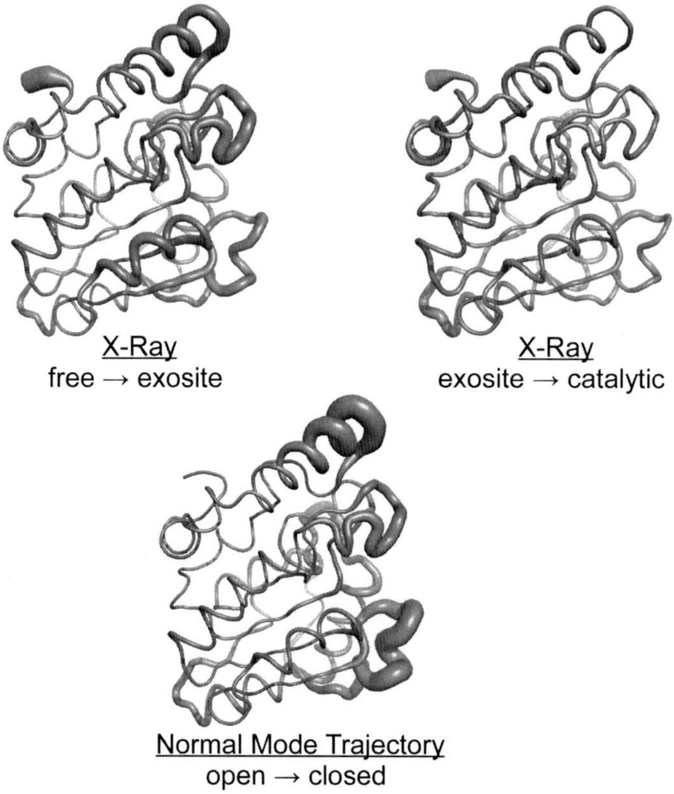

Figure 4. Comparison of the structural perturbations experienced by UNG upon transitioning between the three reaction coordinate conformations for which crystal structures exist. The widths of the lines are proportional to the atom displacements between the indicated structural transition (see text).

The dynamic residues were grouped into five regions of primary sequence and formed the outline of the DNA binding cleft (Figure 3). It is interesting to note that most of the residues in these five regions do not have defined secondary structure in either the free or the bound form, but nevertheless, are relatively rigid in the absence of DNA. The NMR dynamic experiments revealed that these regions experienced a similar rate for conformational exchange (~900 s^{-1}), suggesting that a single global dynamic event modulated the chemical environments for all of these residues.

Implications of UNG Dynamics

While NMR is uniquely able to detect the occurrence of molecular motions, it seldom provides structural details about the interconverting states except in extraordinary situations (*33*, *34*). Meaningful interpretation of dynamic data with respect to structural transitions requires carefully reasoned inferences based upon known structural models. In this study we were fortunate to have high-quality crystallographic structures of three states along the reaction coordinate pathway: free UNG (*35*), UNG bound to non-target DNA with a thymine base in the exosite (*25*), and UNG bound to a fully extrahelical nonreactive substrate analogue (*26*). Since UNG was investigated while bound to non-target DNA, the observed dynamics must be assigned to dynamic fluctuations occurring prior to the formation of the catalytically active complex.

Atomic Coordinates along the Base Flipping Reaction Coordinate

Structural representations for the absolute values of the amide nitrogen displacements of UNG between the three crystallographic states along the base flipping pathway are shown Figure 4, where the thickness of the line is proportional to the magnitude of the atom displacement. A comparison of the displacements upon transitioning from free enzyme to the exosite complex (upper left panel), with the displacements observed in moving from the exosite to specific complex (upper right panel), reveals that the enzyme assumes its catalytically competent conformation almost immediately upon DNA binding. Of interest is the finding that the atomic displacements in moving from the free enzyme to the exosite complex correspond quite well to the dynamic regions of UNG in the DNA bound state (compare Figure 3 and Figure 4). The observed enzymatic dynamics can be reasonably construed as fluctuations between DNA bound states of the enzyme that precede and include the exosite complex.

Normal Mode Analysis

The interpretation of NMR dynamic information is often augmented by computational models describing likely conformational rearrangements. Present computers lack the power to perform Newtonian mechanical simulations over sufficiently long enough time windows to sample the millisecond timescale dynamic events such as those observed with DNA-bound UNG. Monte Carlo sampling strategies are also able to sample different possible protein conformations, but are statistical in nature, and therefore provide no insight into the timescales of the interconversions that might be correlated with experimentally observed dynamics. An alternative approach is the elastic network normal mode analysis (NMA), which treats the protein backbone as simple small blocks interconnected by springs. This method is often able to identify low frequency (and thus low energy) fluctuations in proteins that match observed conformational changes (*36*, *37*).

A normal mode analysis of the UNG structure bound to non-target DNA revealed low frequency normal modes of the enzyme that recapitulated the conformational changes that occur upon DNA binding (compare the bottom and upper left panels in Figure 4). The extrema of the lowest frequency normal mode are also shown in Figure 5 with DNA included for reference. The NMA clearly indicates the presence of a low energy conformational fluctuation that may be best described an open to closed transition.

This open to closed dynamic fluctuation of UNG while bound to DNA is the sort of motion that could potentially allow the enzyme to scan along DNA (the

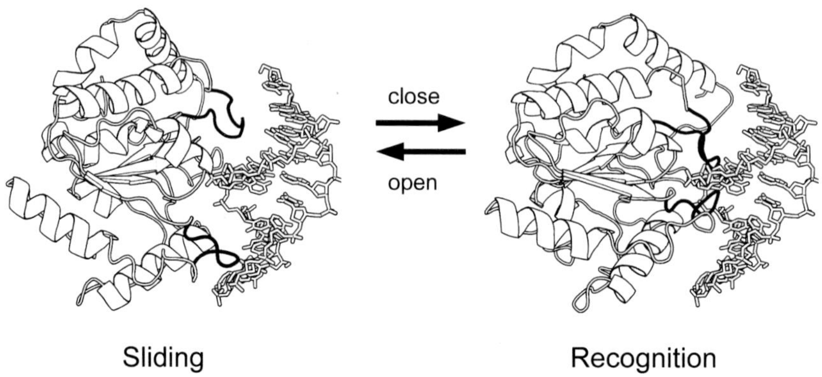

Figure 5. States representing the extrema of the lowest frequency normal modes of the UNG structure bound to non-target DNA. (Computed by the program el Nemo) (38)

open state) and periodically pause to investigate base pair breathing along the DNA strand (the closed state). The observed exchange rate of about 1,000 s^{-1} as determined by NMR dynamic measurements is similar to the estimated lifetime of UNG while interrogating individual base pairs (~0. 5ms). The similarity of these time regimes is consistent with the expectation that the conformational fluctuations of the enzyme must be on the same time scale as its lifetime at each base pair (*14*).

Epilogue

Given the high concentration of non-target DNA binding sites in the genome, and the micromolar binding affinity of UNG for nonspecific DNA sequences (*14*, *31*), UNG is seldom found in a free state *in vivo*. Therefore the functional dynamic behavior of UNG has apparently evolved while it is bound to DNA, providing reasonable justification for why free UNG does not exhibit the same dynamic fluctuations. Importantly, the dynamic motions of UNG have evolved to match the intrinsic breathing motions of thymine and uracil bases so as to optimize the damage search process. We would therefore anticipate that the dynamic behavior of other base excision repair glycosylases would also be matched to the intrinsic dynamics of their respective damaged bases. Although our biophysical understanding of the search and uracil recognition mechanism of UNG has marched considerably forward in the last years, these NMR and structural studies of UNG interacting with naked B DNA represent only a small part of the *in vivo* recognition problem. The future points towards understanding damage recognition and dynamics in the presence of nucleosomes, higher order chromatin structure and in supercoiled DNA substrate.

Acknowledgments

This work was supported by NIH grants GM056834 and GM068626 to J.T.S.

References

1. Fischer, E. *Ber. Dtsch. Chem. Ges.* **1894**, *27*, 2984–2993.
2. Pauling, L. *Nature* **1948**, *161*, 707–709.
3. Perutz, M. *Science* **1963**, *140*, 863–869.
4. Austin, R.; Beeson, K.; Eisenstein, L.; Frauenfelder, H.; Gunsalus, I. *Biochemistry* **1975**, *14*, 5355–5373.
5. Henzler-Wildman, K. A.; Thai, V.; Lei, M.; Ott, M.; Wolf-Watz, M.; Fenn, T.; Pozharski, E.; Wilson, M. A.; Petsko, G. A.; Karplus, M. *Nature* **2007**, *450*, 838–844.
6. Tang, C.; Schwieters, C. D.; Clore, G. M. *Nature* **2007**, *449*, 1078–1082.
7. Weiss, S. *Nat. Struct. Biol.* **2000**, *7*, 724–729.
8. Eisenmesser, E. Z.; Millet, O.; Labeikovsky, W.; Korzhnev, D. M.; Wolf-Watz, M.; Bosco, D. A.; Skalicky, J. J.; Kay, L. E.; Kern, D. *Nature* **2005**, *438*, 117–121.
9. Stivers, J. T.; Jiang, Y. L. *Chem. Rev.* **2003**, *34*, 2729–2759.
10. Sousa, M. M. L.; Krokan, H. E.; Slupphaug, G. *Mol. Aspects Med.* **2007**, *28*, 276–306.
11. Berg, O. G.; Winter, R. B.; von Hippel, P. H. *Biochemistry* **1981**, *20*, 6929–6948.
12. Halford, S. E.; Marko, J. F. *Nucleic Acids Res.* **2004**, *32*, 3040–3052.
13. Berg, H. C. *Random Walks in Biology*; Princeton University Press, 1993.
14. Porecha, R. H.; Stivers, J. T. *Proc. Natl Acad. Sci. U.S.A.* **2008**, *105*, 10791–10796.
15. Gowers, D. M.; Wilson, G. G.; Halford, S. E. *Proc. Natl. Acad. Sci. U.S.A.* **2005**, *102*, 15883–15888.
16. Stanford, N. P.; Szczelkun, M. D.; Marko, J. F.; Halford, S. E. *EMBO J.* **2000**, *19*, 6546–6557.
17. Blainey, P. C.; van Oijen, A. M.; Banerjee, A.; Verdine, G. L.; Xie, X. S. *Proc. Natl. Acad. Sci. U.S.A.* **2006**, *103*, 5752–5757.
18. Berg, O. *Chem. Phys.* **1978**, *31*, 47–57.
19. Stivers, J. T. *Chemistry* **2008**, *14*, 786–793.
20. Leijon, M.; Leroy, J. *Biochimie* **1997**, *79*, 775–779.
21. Moe, J. G.; Russu, I. M. *Biochemistry* **1992**, *31*, 8421–8428.
22. Gueron, M.; Kochoyan, M.; Leroy, J. L. *Nature* **1987**, *328*, 89–92.
23. Cao, C.; Jiang, Y. L.; Stivers, J. T.; Song, F. *Nat. Struct. Mol. Biol.* **2004**, *11*, 1230–1236.
24. Cao, C.; Jiang, Y. L.; Krosky, D. J.; Stivers, J. T. *J. Am. Chem. Soc.* **2006**, *128*, 13034–13035.
25. Parker, J. B.; Bianchet, M. A.; Krosky, D. J.; Friedman, J. I.; Amzel, L. M.; Stivers, J. T. *Nature* **2007**, *449*, 433–437.
26. Parikh, S. S.; Walcher, G.; Jones, G. D.; Slupphaug, G.; Krokan, H. E.; Blackburn, G. M.; Tainer, J. A. *Proc. Natl. Acad. Sci. U.S.A.* **2000**, *97*, 5083–5088.
27. Slupphaug, G.; Mol, C. D.; Kavli, B.; Arvai, A. S.; Krokan, H. E.; Tainer, J. A. *Nature* **1996**, *384*, 87–92.
28. McConnell, H. M. *J. Chem. Phys.* **1958**, *28*, 430–431.

29. Mulder, F. A. A.; Skrynnikov, N. R.; Hon, B.; Dahlquist, F. W.; Kay, L. E. *Biophys. Chem.* **1988**, *17*, 451–479.
30. Cavanagh, J. *Protein NMR Spectroscopy: Principles and Practice*; Academic Press: 2007; pp 243−300.
31. Friedman, J. I.; Majumdar, A.; Stivers, J. T. *Nucleic Acids Res.* **2009**, *37*, 3493–3500.
32. Palmer, A. G., 3rd; Kroenke, C. D.; Loria, J. P. *Methods Enzymol.* **2001**, *339*, 204–238.
33. Skrynnikov, N. R.; Dahlquist, F. W.; Kay, L. E. *J. Am. Chem. Soc.* **2002**, *124*, 12352–12360.
34. Tang, C.; Schwieters, C. D.; Clore, G. M. *Nature* **2007**, *449*, 1078–1082.
35. Mol, C. D.; Arvai, A. S.; Slupphaug, G.; Kavli, B.; Alseth, I.; Krokan, H. E.; Tainer, J. A. *Cell* **1995**, *80*, 869–878.
36. Bahar, I.; Rader, A. *Curr. Opin. Struct. Biol.* **2005**, *15*, 586–592.
37. Dobbins, S. E.; Lesk, V. I.; Sternberg, M. J. E. *Proc. Natl. Acad. Sci. U.S.A.* **2008**, *105*, 10390–10395.
38. Suhre, K.; Sanejouand, Y. H. *Nucleic Acids Res.* **2004**, *32*, W610–W614.

Chapter 5

Biochemical and Structural Domain Analysis of Xeroderma Pigmentosum Complementation Group C Protein

Steven M. Shell and Walter J. Chazin*

Department of Biochemistry, Center in Molecular Toxicology, Vanderbilt University, Nashville, TN 37232
***walter.chazin@vanderbilt.edu**

Nucleotide excision repair (NER) is a DNA repair pathway conserved between prokaryotes and eukaryotes tasked with removing bulky lesions that occur throughout the genome. A unique feature of NER is the ability to recognize a wide variety of chemically and structurally unrelated lesions without the need for multiple lesion-specific proteins. The XPC family proteins are the primary lesion recognition factors in global genome repair NER (GGR) in eukaryotes. GGR is capable of repairing lesions without regard to the transcriptional activity of the damaged DNA and is a powerful guardian of genome integrity. Although XPC has been the subject of intense biochemical investigation, little is known of the molecular mechanism of XPC activity. Currently the structure of the *Saccharomyces cerevisiae* XPC homolog, Rad4, is the only atomic resolution model for damage recognition by XPC. This chapter will review the current knowledgebase for the structural biology of XPC and how determining the structure of Rad4 has influenced the current model of DNA damage recognition in GGR. Specific attention will be placed on comparing and contrasting the structure of Rad4 with the available biochemical data for the human XPC protein.

Introduction

All cells are under constant attack by endogenous and exogenous agents that serve to damage DNA and threaten the integrity of the genome. In humans, DNA damage can lead to mutations and genetic rearrangements, which can contribute to the development and progression of cancer, and in the long-term, can ultimately result in death (*1*, *2*). In order to combat the deleterious effects of DNA damage, cells have developed a complement of biochemical pathways that cooperate to identify and repair DNA lesions as well as coordinate with cellular processes such as replication and apoptosis (*3*, *4*).

Nucleotide excision repair (NER) is a conserved DNA repair pathway found in both prokaryotic and eukaryotic organisms. NER is the primary repair pathway for removing bulky nucleotide lesions that result from exposure to a variety of environmental genotoxic agents, such as ultraviolet (UV) radiation as well as many industrial byproducts (*3*, *4*). In humans, defects in NER lead to the disorder *Xeroderma pigmentosum* (XP), which is characterized by hypersensitivity to UV exposure and predisposition to developing skin cancers (*5*). Analysis of XP-patient derived cell lines led to the identification of seven proteins, XPA-G, defects in which give rise to the XP phenotype (*6*, *7*). The basic NER process involves lesion identification, strand separation, dual incision of the damaged strand, and excision of an oligonucleotide fragment containing the lesion. The resulting gap is then filled by the replicative polymerase Polδ using the undamaged strand as a template, thereby restoring the original sequence. Biochemical analyses have demonstrated this process proceeds via sequential assembly and remodeling of nucleoprotein complexes at the lesion site (*8*). Although there is no homology among the prokaryotic and eukaryotic proteins that perform NER, the overall process is similar.

One of the hallmarks of NER is its ability to recognize and repair a large variety of chemically and structurally unrelated lesions without the necessity of lesion-specific repair proteins (*9*, *10*). This stands in stark contrast to the base excision repair pathway that require extremely specific proteins to repair individual lesions (see Chapters 2-4). The broad substrate specificity observed in NER is derived from the mechanism by which lesions are identified, which can occur by two distinct processes. In transcription-coupled repair (TCR), DNA lesions cause RNA polymerase to stall, eliciting the NER response. However, TCR is limited only to the transcribed strand of actively transcribed regions of the genome (*11*). Global genome repair (GGR) overcomes this limitation by surveying the entire genome independent of RNA polymerase activity. In GGR, the presence of DNA lesions is identified by the activity of the XPC protein (*12*, *13*). XPC is a 940-residue DNA binding protein that was found to correct the XP complementation group C deficiency (*13–15*). Biochemical analysis suggests XPC acts as a DNA helix probe, interrogating the stability of the duplex to discriminate damaged and undamaged DNA (*16*). In most cases the presence of DNA lesions thermodynamically destabilizes the DNA helix (*17*, *18*). It is believed that XPC primarily senses this helix destabilization rather than the lesion itself (*13*, *19*, *20*). This indirect readout model eliminates the need to recognize specific DNA adducts and expands the spectrum of lesions XPC can identify.

XPC has been characterized as a molecular matchmaker in that it marks the lesion site and recruits additional NER proteins, but does not participate in the entire repair process (*6*, *8*). XPC performs this role through its ability to interact with DNA and form complexes with a variety of NER proteins (*20–22*). Although a wealth of biochemical information is available concerning the role of XPC in NER, very little is known of the structural details underlying human XPC activity. This chapter will summarize the current structural and biochemical knowledgebase for XPC in the context of its role in genome maintenance.

XPC Protein Domain Architecture and Structure

The X-ray crystal structure of *Saccharomyces cerevisiae* Rad4 protein is the only atomic resolution structure known for an XPC homolog (*23*). A protease-resistant 531-residue fragment located in the central region of Rad4 containing the minimum DNA binding fragment was expressed and isolated to determine the structure. This fragment corresponds to a conserved region in XPC family proteins that has previously been shown to contain the DNA binding activity of the human XPC (*20*). *In vivo*, Rad4 exists in complex with the Rad23 protein, which is believed to protect Rad4 from ubiquitin-mediated proteasome degradation and stimulate NER activity (*24*), an interaction conserved in the human system (*25*, *26*). Since Rad23 stabilizes Rad4 *in vitro*, Rad4 was expressed and crystallized in complex with a truncated Rad23 protein containing the Rad4 binding domain.

The crystal structure of the Rad4-Rad23 complex shows an extended Rad4 DNA binding apparatus consisting of four domains, an N-terminal transglutaminase-like domain (TGD) and three β-hairpin domains (BHD1-3). The Rad4 TGD contains a core transglutaminase fold homologous to that observed in the crystal structure of peptide-*N*-glycanase (PNGase) (*27*). However, the Rad4 TGD lacks the catalytic triad (cystine-histidine-aspartic acid) characteristic of the transglutaminase superfamily enzymatic activity. Rad23 forms a four-helix saddle structure that binds Rad4 primarily through hydrophobic interactions with the TGD, similar to its interaction with PNGase (*28*). Primary sequence alignment indicates the BHD domains of Rad4 share homology with the oligonucleotide binding (OB) fold found in ssDNA binding proteins, such as replication protein A (RPA) (*29*). However, the structures of BHD1-3 more closely resemble the DNA binding domain of the NER protein XPA than the RPA OB-fold (*23*).

The XPC Transglutaminase-like Domain (TGD)

Multiple sequence alignment among XPC homologs reveals that domain architecture is well conserved among eukaryotes; there is 23% sequence identity between the yeast Rad4 and human XPC proteins. Importantly, Rad4 and XPC display 31% sequence identity in the BHD3 domain, which suggest that the helix probing mechanism is highly conserved (*23*). In contrast, the TGD domain displays a higher degree of variability among XPC proteins in spite of being a conserved motif. Multiple sequence alignment reveals many higher order eukaryotes contain a large insertion that divides the TGD domain into two parts.

The insertion, absent in Rad4, is poorly conserved, has low sequence complexity and has been shown by NMR to be disordered in human XPC (*21*).

Bunick *et al.* proposed that the TGD insertion is dispensable for folding of the domain. Using the crystal structure of the yeast PNGase (PDB ID 1X3W) (*27*) as a template, the authors generated a homology model of the XPC TGD domain in which the insertion was truncated by 150 residues (Figure 1, PDB ID 1CBM) (*21*). The model fully mimics the transglutaminase fold in the PNGase structure and overlays well with the TGD domain in the Rad4 protein. The model shows that the N-terminal (pink) and C-terminal (blue) segments of the TGD domain fold back with the insertion extruded as a long flexible loop (green). Although it is unclear if this protrusion has a function in XPC activity, it is important to note that sequence alignment shows the N- and C-termini of the TGD insertion map to a loop in the Rad4 TGD that participates in dsDNA binding (*23*). Thus, it is likely these residues in human XPC participate in DNA binding.

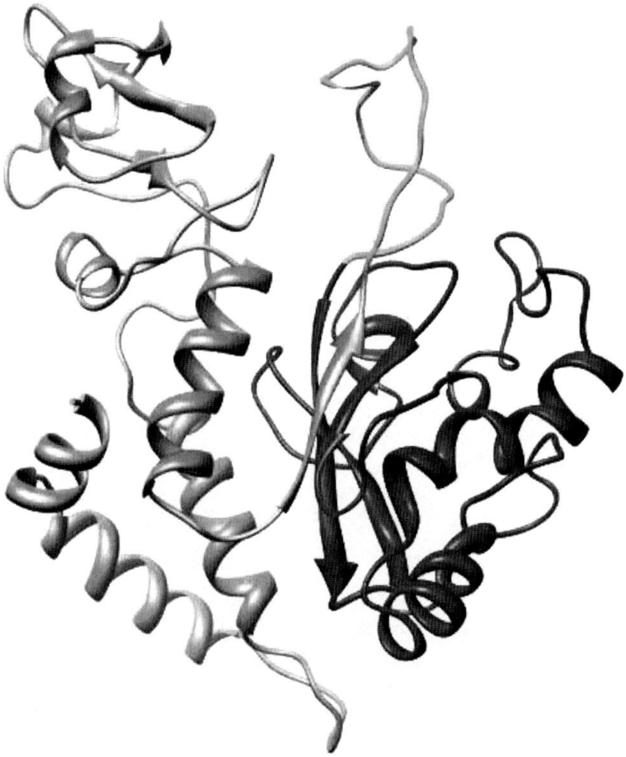

Figure 1. Homology model of the human XPC Transglutaminase-like domain (TGD). The human XPC TGD was modeled using the crystal structure of yeast PNGase (PDB ID: 1X3W) as the template. Human XPC contains a 180-residue insertion between the N-terminal (pink) and C-terminal (blue) segments of the TGD relative to the Rad4 TGD, which was truncated by 150 residues in the model (shown in green). The model demonstrates the two segments of the TGD fold back upon themselves, extruding the insertion as a long flexible loop, to form the tertiary fold of the domain.

The TGD domain has been shown to mediate critical protein-protein interactions, in particular the interaction between XPC and HR23B (human homolog of Rad23) (*20*, *21*, *23*). This model is supported by the crystal structure of the Rad4-Rad23 complex. Rad23 binds to the Rad4 TGD through multiple interactions with helix α2 located in the N-terminus of the domain and helix α13 located in the C-terminus (*23*). Although previously thought to bind in the C-terminal region of XPC (*30*), multiple studies have proposed that HR23B binds XPC on both segments of the TGD. Uchida *et al.* demonstrated that HR23B interacts strongly with a fragment of XPC containing residues 118-940 and significantly weaker with a fragment (residues 496-940) lacking the N-terminal TGD fragment (*20*). Bunick *et al.* also found that HR23B binds to an XPC fragment containing residues 492-940 and stimulated ssDNA binding activity; however, the interaction between HR23B and XPC(492-940) was only transient (*21*). These results suggest HR23B has two contact points; one on the N-terminal and the other in the C-terminal segments of the TGD; which act cooperatively to stabilize the complex.

XPA also binds XPC through interactions with the TGD domain. Previous studies found a direct physical interaction between XPA and the XPC-HR23B complex (*31*, *32*). Bunick *et al.* found XPA binds to both the N-terminal and C-terminal segments of the TGD, but not to the unstructured insertion. XPA exhibited much greater affinity for an N-terminal TGD construct and the interaction was subsequently mapped to a region of the XPC N-terminal TGD containing residues 248-325 (*21*). Given the predicted tertiary structure of the XPC-TGD and evidence of multiple contact points for HR23B and XPA on both the N-terminal and C-terminal segments of the TGD, further investigation is required to determine whether these proteins engage contiguous binding surface(s) on XPC or if multiple discrete binding modules are involved.

Structural Interactions between Damaged DNA and Rad4

The Rad4-Rad23 complex was co-crystallized with a DNA substrate containing a single cyclobutane pyrimidine dimer (CPD) lesion in a three-nucleotide thymidine-thymidine mismatch (to promote complex formation). Figure 2 is a ribbon diagram of the Rad4-Rad23-damaged DNA structure (PDB ID 2QSG) (*23*). This structure reveals Rad4 binds the substrate in two modes. The TGD and BHD-1 domains (shown in gold and pink, respectively) form a dsDNA-binding clamp that contacts the dsDNA region downstream of the lesion, occluding a total of 11 base pairs. Rad23 (which is not shown in Figure 2 for clarity) interacts with a region of the TGD located away from the DNA binding surface and does not directly contribute to DNA binding activity.

BHD2 and BHD3 form a hand-like structure that cooperatively binds primarily to the undamaged strand opposite the CPD lesion occluding a four-base stretch of the substrate (Figure 3). BHD2 (blue) and BHD3 (red) form a groove that contacts with the sugar-phosphate backbone of the undamaged strand. This contact brings the β-hairpin of BHD3 into close proximity to the helix, where it is found inserted between the strands at the lesion site. The two base pairs

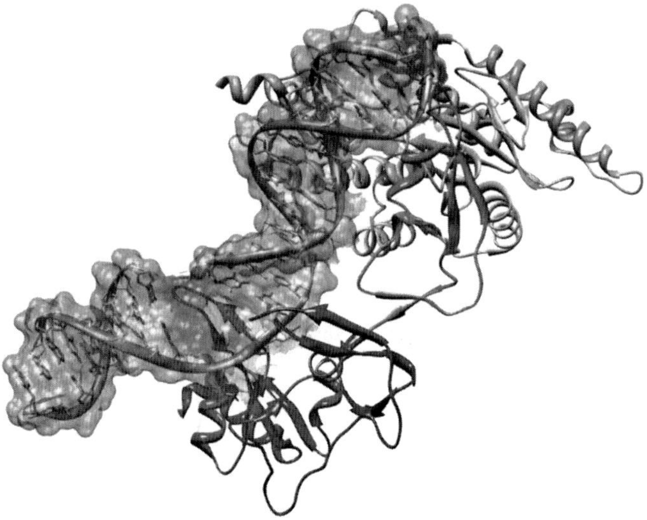

Figure 2. Crystal structure of Rad4 bound to damaged DNA. The Rad4 DNA binding fragment was co-crystallized bound to a 24 base pair duplex substrate containing a cyclobutane pyrimidine dimer (CPD) lesion located in a 2-nucleotide mismatch (PDB ID 2QSF). The TGD (gold) and BHD1 (pink) domains form a double-stranded DNA binding clamp engaging the substrate 3' with respect the to lesion. The BHD2 (blue) and BHD3 (red) domains bind the single-stranded DNA region resulting from thermodynamic destabilization of the helix by the presence of the CPD lesion. The β-hairpin loop of BHD3 inserts between the strands rotating two base pairs out of the duplex. Rad4 does not make any contact with the damaged nucleotides, which are not observed in the crystal structure and are believed to be dynamically disordered. (see color insert)

containing the CDP lesion are rotated out of the base stack enabling the BHD3 β-hairpin to make extensive contacts with the nucleotides flanking the lesion site on both the damaged and undamaged strands. In addition, BHD2 and BHD3 form nucleotide-binding pockets that interact with the flipped out nucleotides on the undamaged strand. The nucleotide-binding pocket appears capable of binding any combination of the four native DNA bases; however, due to size limitations would not accommodate the damaged bases. This inability to bind the damaged nucleotides imparts strand-specificity to the BHD2/3-DNA interaction and may aid in strand discrimination in subsequent NER steps.

Scanning force electron microscopy had previously demonstrated that XPC binding induces DNA bending when the substrate contains a single cholesterol lesion (*33*). This observation is consistent with the crystal structure, which reveals Rad4 induces a 42° bend in the helix axis when in complex with damaged DNA. The bend in the DNA is stabilized by interaction of the BHD3 β-hairpin between the strands as well as by interactions between the flipped out nucleotides and the BHD2 and BHD3 domains. DNA bending may also help stimulate the recruitment of the XPA protein, which binds tightly to deformed DNA structures and is necessary factor for progression of NER (*34*, *35*).

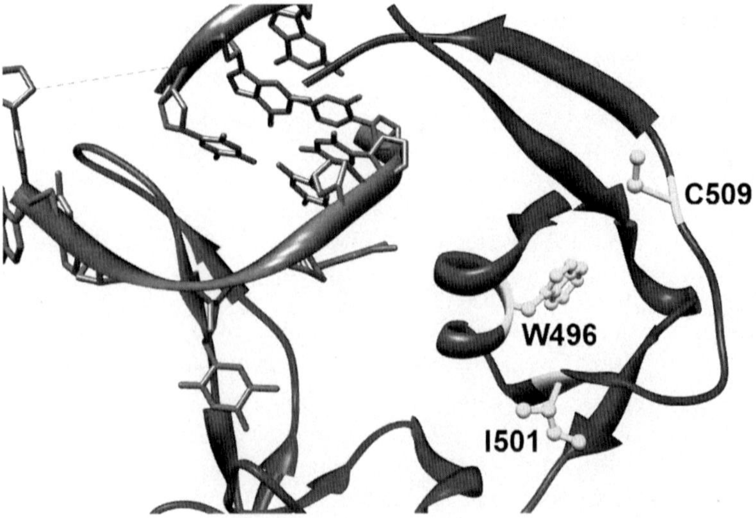

Figure 3. Disposition in Rad4 structure of disease-associated mutations in XPC. Zoomed-in view of Rad4 BHD2 (blue) and BHD3 (red) in contact with CPD-damaged DNA (PDB ID: 2QSF). The β-hairpin of BHD3 inserts between the strands at the damage site, causing two base pairs to rotate out of the duplex. Residues W496, I501, and C509 (in yellow) correspond to disease-associated XPC mutations W690S, V697insV, and P703L, respectively. Of note, W496 is buried within the hydrophobic core of BHD2, although studies have shown W690 in XPC is defective in binding ssDNA.

Comparison of the Rad4 apo- and damaged-DNA-bound structures also illustrates that Rad4 undergoes conformational changes as a result of binding damaged DNA. The four domains of Rad4 are linked by a series of short hinges, giving flexibility to the protein. When aligned on the TGD domain the apo-BHD2/3 domains extend away from the helix while the TGD/BHD1 dsDNA clamp engages the duplex. Conversely, when Rad4 binds damaged DNA the BHD2/3 domains contract toward the helix bringing the BHD3 β-hairpin into close proximity to the base stack. Thus, the Rad4-Rad23 crystal structures suggest an induced-fit mechanism for damaged DNA binding. Presumably the presence of a lesion destabilizes the interactions between the strands thereby lowering the energy penalty for the rotation of the damaged based pairs into an open conformation, allowing invasion of the BHD3 β-hairpin into the helix (*16*). In the absence of helix distortion in an undamaged substrate, the energy required to locally melt the duplex is too great and Rad4 more readily releases from the substrate. In sensing the helix distortion induced by the presence of a lesion, rather than interacting directly with the adduct in a specific manner, Rad4 is capable of identifying a large variety of unrelated lesions provided they induce some degree of thermodynamic instability in the damaged duplex (*36–38*).

Structural Insights into *Xeroderma pigmentosum* Mutations

Currently over 40 individual mutations in the *XPC* gene have been linked to the development of *Xeroderma pigmentosum*, many of which result in early termination of protein synthesis (*39*, *40*). However, four mutations have been identified that result in a defective full-length protein. These mutations are located in the DNA binding region of the XPC protein, and all but one (P334H) can be mapped directly onto the Rad4 crystal structure. Two mutations, V697insV and P703L, map closely together in the hydrophobic core of the Rad4 BHD2 domain (Figure 3). The corresponding Rad4 residues, I502 and C509, respectively, are not directly involved in DNA binding but rather stabilize the domain structure. Although little information is available for the P703L mutant, it is believed both P703L and V697insV lead to loss of DNA binding function (*41*, *42*). Indeed, biochemical analysis demonstrates XPC-V697insV weakens the XPC-damaged DNA binding activity *in vitro* and prevents chromatin association following localized UV irradiation *in vivo*. Surprisingly, in addition to suppressing DNA binding, V697insV also inhibits the XPC interaction with HR23B (*41*). While disruption of this protein-protein interaction leads to increased proteasome-mediated XPC degradation *in vivo* (*25*, *43*), the origin of this effect remains unclear.

W690S is the most intriguing XP mutation. Located in the C-terminal domain associated with damaged DNA binding, multiple reports have demonstrated this mutation decreases the affinity of XPC for ssDNA substrates as well as the ability of XPC to recruit TFIIH to damage sites (*20–22*). Maillard *et al.* have proposed that tryptophan 690 serves as an aromatic sensor within the C-terminal domain of XPC. In this model, strategically aligned aromatic residues stack with the exposed nucleotide bases thereby stabilizing the nucleoprotein complex (*29*). In RPA, the major eukaryotic ssDNA binding protein, the presence of conserved aromatic residues within the OB folds is believed to be essential for ssDNA binding (*44*). However, W496 (the Rad4 analog of W690) is buried within the BHD2 hydrophobic core (Figure 3). Bunick *et al.* reported the XPC-W690S mutation does not alter the susceptibility of the XPC C-terminal domain to limited proteolysis, suggesting the structure remains intact (*21*). Further studies are required in order to determine if W690 is an ssDNA binding residue or alters the conformation of BHD2 in a manner that causes reduced ssDNA binding affinity.

The fourth XPC mutation, P334H, is unique in that it is located in the large insertion in the TGD domain close to the N-terminal end as the insertion exits the TGD. The proline at this position is unique to human XPC and is not analogous to any residues even in the closest mammalian XPC homologues (*45*). Localized UV-irradiation experiments demonstrate that P334H is fully capable of sensing the presence of DNA lesions as evidenced by co-localization in CPD foci. P334H is also capable of initiating the GGR response through recruitment of the TFIIH complex. However, *in vitro* analysis indicates TFIIH activity is reduced in reactions containing purified P334H protein. In addition, XPA recruitment, a necessary step in the progression of GGR, is delayed in P334H transfected cells (*41*). Given neither XPA nor TFIIH have been shown to directly interact with the TGD insertion, it is unclear how the P334H mutation affects the interactions between these factors and XPC and requires further study.

XPC and Base Excision Repair (BER)

Although XPC is generally regarded as an NER factor, it has been shown to interact with the DNA glycosylase repair proteins involved in the BER pathway (*46–48*). *In vitro* DNA incision assays have shown XPC stimulates hydrolysis of oligonucleotide substrates containing an 8-OH-guanine lesion by the Ogg1 glycosylase (*46*). However, Ogg1 activity was unaltered when incubated with full-length XPC containing the P334H mutation. Far-western blot analysis revealed the P334H mutation prevents a protein-protein interaction between the N-terminal of XPC and Ogg1. In addition, XPC-deficient cells or cells expressing the P334H mutant display increased sensitivity to oxidative DNA damaging agents (*41*). These results suggest the XPC-mediated stimulation of BER is a physiologically relevant component of the DNA damage response to oxidative damage.

Although Ogg1 appears to only interact with the N-terminal 474 residues of XPC, this fragment is not sufficient to stimulate BER activity. Cell survival assays demonstrate cells expressing either the P334H mutant protein or a fragment of wild-type XPC containing residues 1-579 demonstrated the same sensitivity to the oxidizing agent $KBrO_3$. In contrast, cells expressing the full-length wild-type XPC were more resistant. This indicates that both the protein-protein interaction between XPC and DNA glycosylases and damaged DNA binding by XPC are critical to the stimulatory role XPC plays in BER (*41*). These results suggest that once XPC identifies the presence of a lesion it may act as a repair pathway switch; recruiting the correct repair proteins necessary to remove the lesion through protein-protein interactions (*49*). However, at present it is unclear if or how XPC could discriminate between lesions repaired by NER or BER.

Discussion

The crystal structure of the yeast Rad4 protein has provided new insight and perspective relating to the design and interpretation of cellular and biochemical data collected for human XPC. Recently the Rad4 structure was used to design novel truncation mutants to dissect the damage response activity of XPC *in vivo* (*50*). Camenisch *et al.* found that BHD3 is not required for XPC to co-localize with damaged DNA in UV-irradiated cells, but is required for retention at the damage site. Using the Rad4 structure as a guide (*23*), a 25-residue β-extension of BHD2 was identified as the minimum necessary feature to differentiate between damaged and undamaged DNA in cells (*50*). The conserved residues in Rad4 form a nucleotide binding pocket that traps one of the two bases flipped out by insertion of the BHD3 β-hairpin between the two strands at the damage site (*23*). However, the β-extension interaction was insufficient to form a stable nucleoprotein complex, which requires insertion of the BHD3 β-hairpin (*50*). The authors therefore proposed a two-step model for sensing damaged DNA.

First XPC quickly scans the duplex for the presence of extrahelical nucleotides using the BHD2 β-extension. When unpaired nucleotides are encountered, a transient intermediate complex with the substrate is formed that brings the BHD3 domain in close proximity to the duplex. Second, BHD3 thoroughly interrogates the stability of the duplex (*50*). As discussed above, only when a DNA lesion is present is the energetic cost of inserting the BHD3 β-hairpin between the strands favorable, which results in a stable nucleoprotein complex capable of recruiting downstream repair proteins (*23*, *50*). Although this is an attractive model for explaining how XPC can rapidly identify lesions throughout large eukaryotic genomes, it has yet to be verified structurally.

Conspicuously absent from the available biochemical data for human XPC is a defined role for the TGD in NER. The TGD has been shown to be critical for NER *in vivo*, as truncated XPC constructs lacking the N-terminal structured domain of the TGD are only minimally active in NER (*20*, *50*). The Rad4 crystal structure clearly illustrates the TGD plays an integral role in binding the undamaged dsDNA 3′ to the lesion, with all DNA contacts located in the N-terminal segment of the domain (*23*). This is in stark contrast to the biochemically-defined minimum XPC fragment necessary for stable binding to damaged DNA, which includes residues 607-831 and is primarily composed of the three β-hairpin domains (*20*, *50*). Therefore, what is the role of the TGD? In both XPC and Rad4 the TGD serves as a protein interaction module, most notably with Rad23 and XPA (*15*, *21*, *22*, *24*, *26*, *30–32*). While the XPA interacting domain has been mapped for XPC, the XPC binding domain(s) for XPA have yet to be determined nor is it known whether XPA binds the two TGD segments simultaneously or independently.

Although the role of the XPC:XPA interaction in NER is unknown, a recent study suggests it may play a role in the damage verification process. Sugasawa *et al.* have proposed that XPC, XPA, and the TFIIH core form a ternary complex capable of translocation along the duplex, with lesion verification occurring when the XPD helicase stalls after encountering a chemically modified nucleotide (*51*). However, it is not readily apparent from the Rad4-damaged DNA complex how translocation would occur. Additionally, it has been reported that XPA, in conjunction with RPA, causes the dissociation of XPC from the damage site (*6*, *8*). While it is possible the XPC:XPA interaction induces remodeling of the XPC architecture to promote the transitions from recognition to verification to dissociation from the substrate, currently there is no structural evidence to support this conclusion.

While the crystal structure of the yeast Rad4 protein bound to damaged DNA has provided a significant step forward in our understanding of lesion identification in NER, clearly many questions remain and some new ones have emerged. Although Rad4 is homologous to human XPC significant differences exist between the two proteins, including substantial insertions and deletions in the DNA binding domains. Moreover, analysis of the Rad4 structure does not adequately address question regarding the role of point mutations, such as the W690S and P334H, in XP disorder in humans and how they affect XPC function. Also, currently little is known about the structural basis for the multiple protein-protein interactions of XPC in both NER and BER. Consequently, continued investigation of the structural and biochemical basis for function is imperative to better understand the activity of human XPC. XPC plays a critical role in the eukaryotic response to DNA damage arising from a variety of genomic insults. As such, understanding the structural details underlying its mechanisms of action will provide invaluable insight into human health and susceptibility to environmental carcinogens. Also, given many chemotherapeutics act through inducing DNA damage repaired by the NER pathway (*52, 53*), elucidating the precise interactions involved in DNA lesion identification and subsequent protein complexes formed have the potential to facilitate the development of novel anti-cancer treatments.

Acknowledgments

The authors would like to thank Dr. Seth Chitayat for critical reading of this manuscript. NER research in the Chazin laboratory is supported by operating grants (RO1 GM065484 and PO1 CA092584) and training grant support to SMS (T32 ES007028) from the National Institutes of Health.

References

1. Braithwaite, E.; Wu, X.; Wang, Z. *Mutat. Res.* **1999**, *424*, 207–219.
2. Marrot, L.; Meunier, J. R. *J. Am. Acad. Dermatol.* **2008**, *58*, S139–148.
3. Sancar, A. *Annu. Rev. Biochem.* **1996**, *65*, 43–81.
4. Sancar, A.; Lindsey-Boltz, L. A.; Unsal-Kacmaz, K.; Linn, S. *Annu. Rev. Biochem.* **2004**, *73*, 39–85.
5. Lehmann, A. R. *Biochimie* **2003**, *85*, 1101–1111.
6. Riedl, T.; Hanaoka, F.; Egly, J. M. *EMBO J.* **2003**, *22*, 5293–5303.
7. Sancar, A.; Hearst, J. E. *Science* **1993**, *259*, 1415–1420.
8. Park, C. J.; Choi, B. S. *FEBS J.* **2006**, *273*, 1600–1608.
9. Naegeli, H. *FASEB J.* **1995**, *9*, 1043–1050.
10. Wood, R. D. *Annu. Rev. Biochem.* **1996**, *65*, 135–167.
11. Fousteri, M.; Mullenders, L. H. *Cell Res.* **2008**, *18*, 73–84.
12. Sugasawa, K.; Ng, J. M.; Masutani, C.; Iwai, S.; van der Spek, P. J.; Eker, A. P.; Hanaoka, F.; Bootsma, D.; Hoeijmakers, J. H. *Mol. Cell* **1998**, *2*, 223–232.
13. Batty, D.; Rapic'-Otrin, V.; Levine, A. S.; Wood, R. D. *J. Mol. Biol.* **2000**, *300*, 275–290.
14. Masutani, C.; Sugasawa, K.; Yanagisawa, J.; Sonoyama, T.; Ui, M.; Enomoto, T.; Takio, K.; Tanaka, K.; van der Spek, P. J.; Bootsma, D.; et al. *EMBO J.* **1994**, *13*, 1831–1843.
15. Reardon, J. T.; Mu, D.; Sancar, A. *J. Biol. Chem.* **1996**, *271*, 19451–19456.
16. Geacintov, N. E.; Broyde, S.; Buterin, T.; Naegeli, H.; Wu, M.; Yan, S.; Patel, D. J. *Biopolymers* **2002**, *65*, 202–210.
17. Cheng, X.; Kelso, C.; Hornak, V.; de los Santos, C.; Grollman, A. P.; Simmerling, C. *J. Am. Chem. Soc.* **2005**, *127*, 13906–13918.
18. Meneni, S. R.; Shell, S. M.; Gao, L.; Jurecka, P.; Lee, W.; Sponer, J.; Zou, Y.; Chiarelli, M. P.; Cho, B. P. *Biochemistry* **2007**, *46*, 11263–11278.
19. Trego, K. S.; Turchi, J. J. *Biochemistry* **2006**, *45*, 1961–1969.

20. Uchida, A.; Sugasawa, K.; Masutani, C.; Dohmae, N.; Araki, M.; Yokoi, M.; Ohkuma, Y.; Hanaoka, F. *DNA Repair (Amst)* **2002**, *1*, 449–461.
21. Bunick, C. G.; Miller, M. R.; Fuller, B. E.; Fanning, E.; Chazin, W. J. *Biochemistry* **2006**, *45*, 14965–14979.
22. Yokoi, M.; Masutani, C.; Maekawa, T.; Sugasawa, K.; Ohkuma, Y.; Hanaoka, F. *J. Biol. Chem.* **2000**, *275*, 9870–9875.
23. Min, J. H.; Pavletich, N. P. *Nature* **2007**, *449*, 570–575.
24. Madura, K. *Trends Biochem. Sci.* **2004**, *29*, 637–640.
25. Ng, J. M.; Vermeulen, W.; van der Horst, G. T.; Bergink, S.; Sugasawa, K.; Vrieling, H.; Hoeijmakers, J. H. *Genes Dev.* **2003**, *17*, 1630–1645.
26. Sugasawa, K.; Masutani, C.; Uchida, A.; Maekawa, T.; van der Spek, P. J.; Bootsma, D.; Hoeijmakers, J. H.; Hanaoka, F. *Mol. Cell Biol.* **1996**, *16*, 4852–4861.
27. Lee, J. H.; Choi, J. M.; Lee, C.; Yi, K. J.; Cho, Y. *Proc. Natl. Acad. Sci. U.S.A.* **2005**, *102*, 9144–9149.
28. Kim, I.; Ahn, J.; Liu, C.; Tanabe, K.; Apodaca, J.; Suzuki, T.; Rao, H. *J. Cell Biol.* **2006**, *172*, 211–219.
29. Maillard, O.; Solyom, S.; Naegeli, H. *PLoS Biol.* **2007**, *5*, e79.
30. Li, L.; Lu, X.; Peterson, C.; Legerski, R. *Mutat. Res.* **1997**, *383*, 197–203.
31. Thoma, B. S.; Wakasugi, M.; Christensen, J.; Reddy, M. C.; Vasquez, K. M. *Nucleic Acids Res.* **2005**, *33*, 2993–3001.
32. You, J. S.; Wang, M.; Lee, S. H. *J. Biol. Chem.* **2003**, *278*, 7476–7485.
33. Janicijevic, A.; Sugasawa, K.; Shimizu, Y.; Hanaoka, F.; Wijgers, N.; Djurica, M.; Hoeijmakers, J. H.; Wyman, C. *DNA Repair (Amst)* **2003**, *2*, 325–336.
34. Camenisch, U.; Dip, R.; Schumacher, S. B.; Schuler, B.; Naegeli, H. *Nat. Struct. Mol. Biol.* **2006**, *13*, 278–284.
35. Yang, Z.; Roginskaya, M.; Colis, L. C.; Basu, A. K.; Shell, S. M.; Liu, Y.; Musich, P. R.; Harris, C. M.; Harris, T. M.; Zou, Y. *Biochemistry* **2006**, *45*, 15921–15930.
36. Buterin, T.; Meyer, C.; Giese, B.; Naegeli, H. *Chem. Biol.* **2005**, *12*, 913–922.
37. Dip, R.; Camenisch, U.; Naegeli, H. *DNA Repair (Amst)* **2004**, *3*, 1409–1423.
38. Maillard, O.; Camenisch, U.; Blagoev, K. B.; Naegeli, H. *Mutat. Res.* **2008**, *658*, 271–286.
39. Cleaver, J. E.; Thompson, L. H.; Richardson, A. S.; States, J. C. *Hum. Mutat.* **1999**, *14*, 9–22.
40. Chavanne, F.; Broughton, B. C.; Pietra, D.; Nardo, T.; Browitt, A.; Lehmann, A. R.; Stefanini, M. *Cancer Res.* **2000**, *60*, 1974–1982.
41. Bernardes de Jesus, B. M.; Bjoras, M.; Coin, F.; Egly, J. M. *Mol. Cell Biol.* **2008**, *28*, 7225–7235.
42. Jacobelli, S.; Soufir, N.; Lacapere, J. J.; Regnier, S.; Bourillon, A.; Grandchamp, B.; Hetet, G.; Pham, D.; Palangie, A.; Avril, M. F.; Dupin, N.; Sarasin, A.; Gorin, I. *Br. J. Dermatol.* **2008**, *159*, 968–973.
43. Xie, Z.; Liu, S.; Zhang, Y.; Wang, Z. *Nucleic Acids Res.* **2004**, *32*, 5981–5990.
44. Bochkarev, A.; Bochkareva, E. *Curr. Opin. Struct. Biol.* **2004**, *14*, 36–42.

45. Rivera-Begeman, A.; McDaniel, L. D.; Schultz, R. A.; Friedberg, E. C. *DNA Repair (Amst)* **2007**, *6*, 100–114.
46. D'Errico, M.; Parlanti, E.; Teson, M.; de Jesus, B. M.; Degan, P.; Calcagnile, A.; Jaruga, P.; Bjoras, M.; Crescenzi, M.; Pedrini, A. M.; Egly, J. M.; Zambruno, G.; Stefanini, M.; Dizdaroglu, M.; Dogliotti, E. *EMBO J.* **2006**, *25*, 4305–4315.
47. Miao, F.; Bouziane, M.; Dammann, R.; Masutani, C.; Hanaoka, F.; Pfeifer, G.; O'Connor, T. R. *J. Biol. Chem.* **2000**, *275*, 28433–28438.
48. Shimizu, Y.; Iwai, S.; Hanaoka, F.; Sugasawa, K. *EMBO J.* **2003**, *22*, 164–173.
49. Isaacs, R. J.; Spielmann, H. P. *DNA Repair (Amst)* **2004**, *3*, 455–464.
50. Camenisch, U.; Trautlein, D.; Clement, F. C.; Fei, J.; Leitenstorfer, A.; Ferrando-May, E.; Naegeli, H. *EMBO J.* **2009**, *28*, 2387–2399.
51. Sugasawa, K.; Akagi, J.; Nishi, R.; Iwai, S.; Hanaoka, F. *Mol. Cell* **2009**, *36*, 642–653.
52. Muggia, F. M.; Fojo, T. *J. Chemother.* **2004**, *16* (Suppl. 4), 77–82.
53. Wozniak, K.; Blasiak *J. Acta Biochim. Pol.* **2002**, *49*, 583–596.

Chapter 6

Effects of Sequence Context on O^6-Alkylguanine DNA Alkyltransferase Repair of O^6-Alkyl-Deoxyguanosine Adducts

Rebecca Guza,[1] Anthony E. Pegg,[2] and Natalia Tretyakova[1,*]

[1]Department of Medicinal Chemistry and the Cancer Center, University of Minnesota, Minneapolis, MN 55455
[2]Department of Cellular and Molecular Physiology, Pennsylvania State University College of Medicine, Hershey, PA 17033
***trety001@umn.edu**

The DNA repair protein, O^6-alkylguanine DNA-alkyltransferase (AGT), specifically recognizes and removes O^6-alkyl substituents on guanine, restoring normal guanine and preventing mutagenesis. When AGT binds O^6-alkyl-deoxyguanosine (O^6-alkyl-dG) containing DNA, the protein interacts with the 1-, N^2-, and 7- positions of the guanine lesion. The repair of O^6-alkyl-dG lesions is a multi-step process. Following AGT binding to DNA, the adducted nucleotide is flipped into the active site of the protein, and the O^6-alkyl substituent is displaced via an S_N2-type mechanism, restoring normal guanine and producing alkylated protein. The rates of AGT repair are influenced by DNA sequence context, secondary structure, and alkyl group identity. The relative rates of AGT-mediated repair of O^6-alkyl-dG lesions are benzyl > methyl > ethyl >> 2-hydroxyethyl >4-(3-pyridyl)-4-oxobutyl. The differences in rates of repair between different alkyl groups and different sequence contexts are not a result of difference in AGT binding and kinetics of nucleotide flipping, since these reaction steps are very fast and are unaffected by DNA sequence. The rate of alkyl transfer is the slowest forward step in the repair of O^6-Me-dG and appears to be dependent on the alkyl group identity and is influenced by the local sequence context. AGT repair of O^6-alkyl-dG lesions is essential for maintenance of genome integrity, and slow repair of these lesions in specific DNA sequences may contribute to the mutational spectra observed in human cancer.

Introduction

Alkylating agents from endogenous and exogenous sources, environmental toxins, and chemotherapeutic agents can react with DNA, resulting in the formation of O^6-alkyl-deoxyguanosine (O^6-alkyl-dG) lesions. These lesions are mutagenic, carcinogenic, and cytotoxic (*1–4*). O^6-alkyl-dG lesions are strongly mispairing, because the alkylation of the O^6 position of guanine alters its ability to form Watson-Crick base pair. Instead of its normal partner (cytosine), O^6-alkyl-dG preferentially pairs with thymine, resulting in G → A transitions (*5*).

Major targets of DNA adduct formation and mutations in smoking-induced lung cancer are the *K-ras* proto-oncogene and the *p53* tumor suppressor gene. Mutations in these critical genes can result in a loss of control over cell proliferation, thereby leading to tumor formation. Interestingly, these mutations are not distributed randomly. The G → A transitions that occur in *K-ras* codon 12 have been shown to correlate with lung cancer incidence, with these mutations observed more frequently in smokers than in non-smokers (*6*). Furthermore, the G → A mutations observed in *K-ras* codon 12 are also found in mouse lung tumors induced by the tobacco carcinogen, NNK (*7*), making NNK a likely causative agent for these genetic changes. NNK induces O^6-methylguanine and O^6-4-(3-pyridyl)-4-oxobutyl -guanine lesions (*7*). Smoking-induced mutations in the *p53* tumor suppressor gene occur in exons 5-8 at endogenously methylated CG dinucleotides, e.g. codons 157, 158, 245, 248, 249, and 273 (*8*). Among these, 18-24% are G → A transition mutations likely induced by O^6-alkylguanine lesions (*8–10*).

A specialized repair protein, O^6-alkylguanine-DNA alkyltransferase (AGT), recognizes and removes the alkyl group from O^6-alkyl-deoxyguanosine lesions in DNA. Wolf *et al.* examined the relationship between the inactivation of the AGT gene by promoter hypermethylation and the mutational spectrum of the *p53* tumor suppressor gene in non-small cell lung cancer (*9*). These authors found that only 8% of lung tumors had G→A transition mutations in the *p53* gene when the promoter region of the gene coding for AGT was not methylated, thereby allowing protein expression (*9*). In contrast, 33% of tumors with inactivated AGT had G→A mutations within the *p53* gene (*9*). Other human cancers that have lower expression levels of AGT have an increased frequency of G→A mutations in the *K-ras* proto-oncogene and the *p53* tumor suppressor gene as compared to tumors with normal expression of AGT (*11–13*). Therefore, AGT appears to play a key role in protection of the genome against G→A transition mutations.

Many structurally related guanine lesions can be repaired by AGT, e.g. O^6-methyl-deoxyguanosine (O^6-Me-dG), O^6-ethyl-deoxyguanosine (O^6-Et-dG), O^6-butyl-deoxyguanosine (O^6-Bu-dG), O^6-2-hydroxyethyl-deoxyguanosine (O^6-He-dG), O^6-4-(3-pyridyl)-4-oxobutyl-deoxyguanosine (O^6-POB-dG), and O^6-benzyl-deoxyguanosine (O^6-Bz-dG) (Figure 1). AGT mediated repair of O^6-Me-dG can be viewed as a multi-step process (Figure 2) (*14*). Following AGT binding to DNA, the adducted nucleotide is flipped into the active site of the protein, and the alkyl group is transferred from the O^6-position of guanine to the active site cysteine of AGT (Cys 145 of the human protein). Finally, the now alkylated AGT dissociates from repaired DNA. Several studies have shown that DNA sequence context can affect the rates of AGT-mediated dealkylation of O^6-alkyl-dG (*15–19*), while others observed little sequence specificity (*20–23*). Here we will review a selection of articles that have explored the effect of sequence surrounding various O^6-alkyl-dG lesions on AGT repair efficiency.

Structure of the AGT Protein

Several X-ray crystal structures of the human AGT protein (hAGT) are now available (*24–27*) (Figure 3). Studies by Daniels *et al.* and Wibley *et al.* have determined the structure of human AGT protein in the absence of DNA and showed that the protein has a two-domain α/β fold structure (*24, 25*). The N-terminal domain contains a three-stranded anti-parallel β-sheet, followed by two helices and is joined to the C-terminal domain by another helix (*24, 25*). The C-terminal domain is made up of a two-stranded parallel β-sheet and five helices (*24, 25*). The C-terminal domain also contains the conserved active site cysteine (Cys 145), the O^6-alkyl-dG binding channel, and a helix-turn-helix (HTH) DNA-binding motif (*24, 25*).

Interestingly, the structure reported by Daniels *et al.* revealed a tetrahedral zinc(II) ion binding site in the N-terminal domain between Cys5, Cys24, His29, and His85 (*25*). A comparison of the X-ray crystal structure with zinc bound to the protein with the X-ray crystal structure without zinc revealed that in the absence of the zinc(II) ion, local distortions in the N-terminal domain are observed, suggesting that the zinc(II) ion facilitates domain interactions and stabilizes protein fold (*25*). Although the zinc(II) ion is ~20Å from the active site cysteine, the presence of zinc also lowers the pKa of the Cys145 and increases its nucleophilicity (*28, 29*).

Structures of hAGT bound to double stranded DNA revealed that the second (recognition) helix in the HTH motif binds within the minor groove of DNA, while the first helix interacts with the phosphodiester backbone (*26, 27*). HTH domain proteins typically bind B-DNA in the major groove, which allows base-specific hydrogen bonds to form and therefore may be important for sequence-specific recognition of DNA. In contrast, AGT binds DNA in the minor groove, leading to a hypothesis that this may be advantageous for sequence-independent binding (*26*).

Crystal structures revealed that a single molecule of AGT occupies ~7 base pairs of the DNA duplex (*26, 27*). The structure of the AGT protein is unaffected by its binding to DNA, but the DNA structure is altered upon protein binding. Specifically, the minor groove of DNA widens by > 3Å, and DNA bends ~15° away from the protein (*26, 27*). Furthermore, the adducted nucleotide is flipped out of the DNA base stack to enter the protein active site, while Arg128 is positioned at the beginning of the recognition helix and takes the place of the

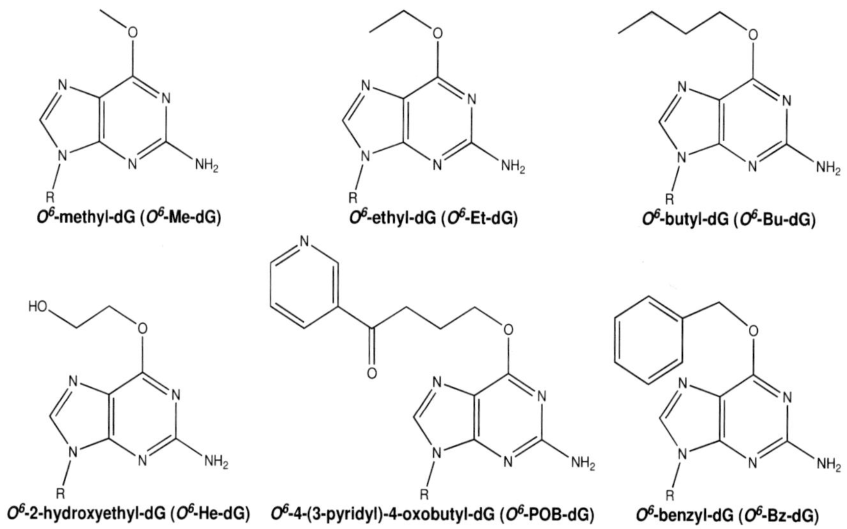

Figure 1. Examples of O^6-alkyl-dG lesions that are recognized and repaired by AGT. R = dG

adducted nucleotide in the DNA duplex (Figure 3) (*26*, *27*). These structural changes are important for the AGT mediated alkyl transfer reaction.

Mechanism of AGT Repair

Once the adducted nucleotide is bound within the active site of the AGT protein, the O^6-substituent is transferred to the active site cysteine (C145) in an irreversible and stoichiometric reaction (*30*). Recognition of O^6-alkylated guanine by the protein is facilitated by hydrogen bond formation between the hydroxyl of Tyr114 and the N3 of guanine (*26*, *27*). Spratt *et al.* showed that AGT binding involves the 1-, N^2-, and 7-positions of guanine, while methyl transfer involves the 3- and O^6-positions of the nucleobase (*31*). The conformation of the protein bound to DNA aligns Cys145 thiol with the O^6-substituent of guanine for in-line displacement (*26*, *27*), facilitating alkyl transfer. A hydrogen bond network around the active site involving His146, Glu172, and a water molecule promotes the deprotonation of the active site cysteine, which generates a highly reactive thiolate anion at Cys145 (*25*, *26*). The thiolate anion acts as a nucleophile, displacing the O^6 substituent of O^6-alkyl-dG and regenerating normal guanine (Figure 4) (*25*, *26*). Recent density functional theory calculations have shown that this proposed mechanism of alkyl transfer is energetically plausible (*32*).

Coulter *et al.* proposed that the local DNA sequence affects the orientation of the O^6-alkyl-dG lesion in the AGT binding pocket and influences the alignment of the O^6-substituent with the active site cysteine, potentially affecting the rate of alkyl transfer (*33*). Depending on sequence context, AGT may bind the lesion in a reactive or a less-reactive conformation (*33*). For examples, O^6-Me-dG and O^6-Et-dG can assume a *syn*-conformation where the O^6-substituent is *syn* with respect to the N1-position of guanine and the substituent points into the helix, or an

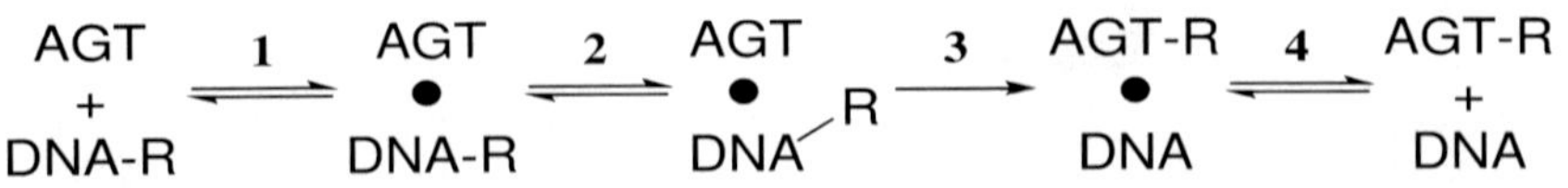

Figure 2. Multi-step process of AGT mediated repair in the removal of an alkyl group from an O^6-alkyl-dG lesion (14). (1) AGT binds to DNA, (2) O^6-alkyl-dG is flipped into the active site of AGT, (3) O^6-alkyl group is transferred from DNA to protein, (4) AGT-DNA complex dissociates. Reproduced with permission. © 2005 The American Society for Biochemistry and Molecular Biology.

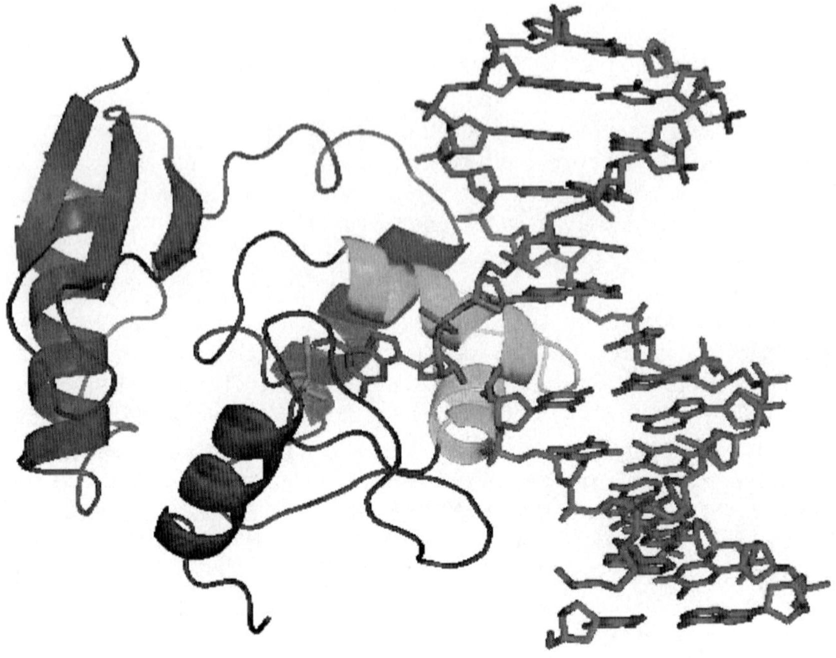

Figure 3. AGT (blue) bound to O^6-Me-dG containing DNA (pink). AGT binds DNA through a helix-turn-helix (HTH) motif (green). The adducted nucleotide is flipped into the active site of the protein where Cys145 (orange) is positioned for S_N2-type removal of the O^6-alkyl group of guanine. (Protein Data Bank accession code 1T38; (26)) (see color insert)

anti-conformation where the O^6-substituent is *anti* with respect to the N1-position of guanine, while the substituent points into the major groove (*33–35*). If the *anti* orientation of the O^6-alkyl group is maintained when the lesion is flipped into the active site of the protein, it is proposed to decrease the rate of AGT repair (*33–35*).

Once the AGT protein is alkylated, it is inactive and is rapidly degraded by the ubiquitin proteolytic pathway (*36–38*). The alkylated AGT undergoes a conformational change resulting from disruption of a hydrogen-bond network in the protein and a steric clash between the alkyl-cysteine and Met134 (*25*, *39*). This conformational change may facilitate the release of the repaired DNA and mediate the *in vivo* detection of the alkylated protein (*25*).

DNA Sequence Effects on the Kinetics of AGT-Mediated Repair

AGT mediated repair of O^6-alkyl-dG is essential for the maintenance of genome integrity. If not repaired prior to DNA replication, O^6-alkyl-dG adducts are misread by DNA polymerases, leading to heritable mutations. Slow repair of O^6-alkyl-dG lesions at specific sites within the genome can lead to mutational "hotspots". As discussed above, inactivation of the AGT gene by hypermethylation of the promoter region results in a significant increase in G → A transition mutations in the *p53* gene of non-small cell lung cancer (*9*). A number of studies examined the effects of local sequence context on repair of O^6-alkyl-dG lesions by human AGT protein. Investigation of AGT-mediated repair has been analyzed using various kinetic conditions. AGT-mediated repair of an O^6-alkyl-dG lesion has been shown to occur with second-order kinetics, as explained by a simple bimolecular reaction (Tables 1-5). Alternatively, the reaction can be studied by first-order kinetics (Table 6). This is explained by the fact that AGT initially forms a complex with DNA that exhibits saturation kinetics, and then the alkyl group is transferred (*40*).

The rate of O^6-alkyl-dG repair by AGT can be affected by the length of the oligonucleotide substrate, as well as lesions location within the sequence. In double stranded DNA oligonucleotides, the repair of O^6-Me-dG by AGT was faster in a 12-mer than in a 6-mer (Table 1) (*41*, *42*). In single stranded oligonucleotides containing O^6-Me-dG, maximal rate of repair was achieved with a 7-mer (Table 2) (*15*). In single stranded oligonucleotides containing an O^6-Bz-dG lesion, the rate of AGT repair increased with oligonucleotide length, but reached a maximal rate when a 5-9-mer was used (Table 3) (*43*). However, AGT mediated repair of O^6-Me-dG and O^6-Bz-dG in single stranded DNA occurred more readily when the O^6-Me-dG lesion was at the 5′ end of the oligonucleotide as opposed to the 3′ end (*19*, *26*), suggesting a directional bias for AGT repair. Daniels *et al.* proposed that AGT molecules are recruited to the 5′ side of the initial DNA-AGT complex faster than to the 3′ side (*26*). AGT has been shown to localize to sites of active transcription *in vivo* (*44*). This directional bias may allow AGT to remain slightly ahead of DNA polymerases, thereby maximizing the efficiency of O^6-alkyl-dG repair (*45*).

Early studies using extracts from HT29 cells that contained human AGT found that the second order rate of demethylation of O^6-Me-dG was greater for a duplex dodecamer when C was 5′ to O^6-Me-dG as compared to G (Table 1) (*16*). Our laboratory employed purified human protein to show that the rate of repair of O^6-Me-dG when it is located at the first G within *K-ras* codon 12 was greater than when it was the second G, although the difference was not statistically significant (*21*). Sequence dependent repair of O^6-Me-dG within *ras* gene sequences was also observed in other studies, in which the rate of repair of O^6-Me-dG by AGT was examined using first order kinetics (Table 6) (*33*, *40*). When O^6-Me-dG was placed in the context of *H-ras* codon 12, the rate of repair was greater when O^6-Me-dG was the first G in the run as opposed to the second G in the run (Table 6) (*33*, *40*). This result in consistent with the observation of GGT → GAT transitions in smoking and NNK-induced tumors (*6*, *7*).

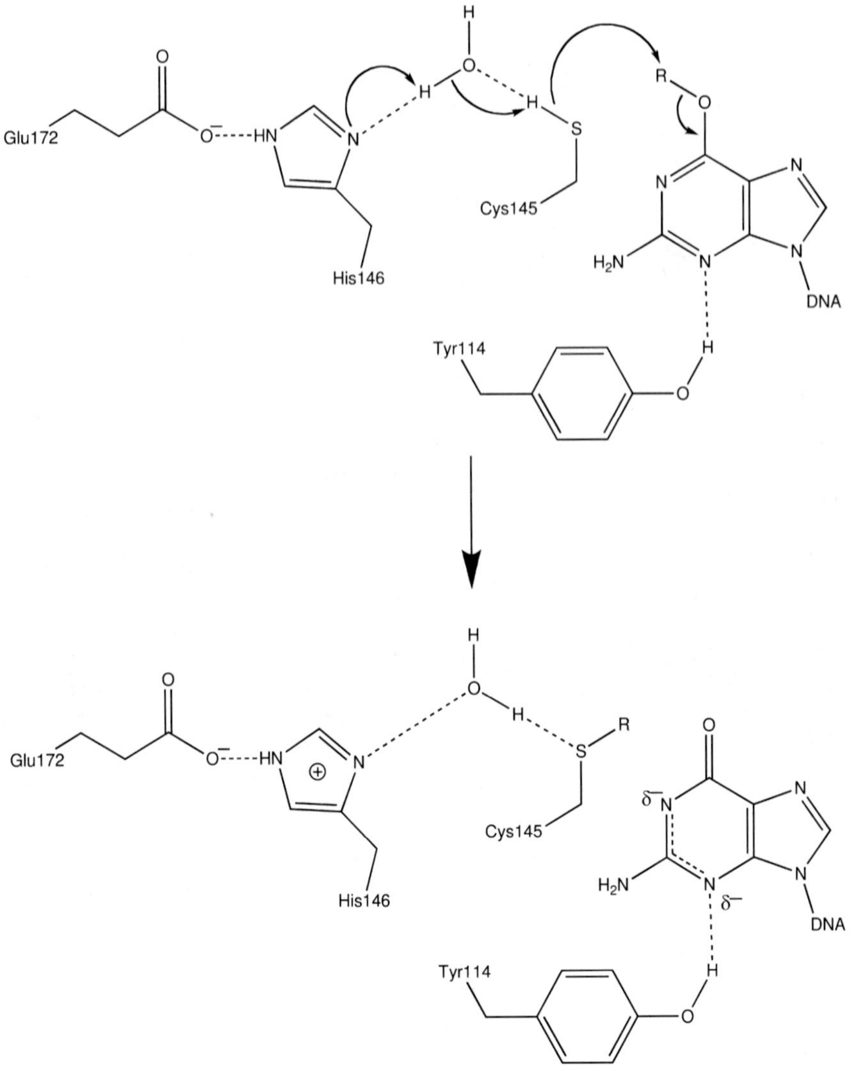

Figure 4. AGT mediated removal of an O[6]*-alkyl lesion from guanine. A hydrogen bond network around the active site promotes the deprotonation of the active site cysteine, which generates a highly reactive thiolate anion at Cys145. The thiolate anion acts as a nucleophile, displacing the O*[6] *substituent of O*[6]*-alkyl-dG and regenerating a normal guanine. Reproduced with permission from Macmillan Publishers Ltd: Nat. Struct. Mol. Biol., Daniels, D. S., Woo, et al. 11, 714-720. © 2004 (26).*

Table 1. Repair of *O*^{6}-Me-dG lesions in double stranded DNA

5′-ACCCGCGTCC[O^6-Me-G]CGCCATGGCC-3′ 3′-TGGGCGCAGG C GCGGTACCGG-5′	$k = 8.8 \pm 0.8 \times 10^6$ M^{-1} s^{-1}	*(53)*
5′-ACCCGCGTC^{Me}C[O^6-Me-G]CGCCATGGCC-3′ 3′-TGGGCGCAG G C GCGGTACCGG-5′	$k = 10.2 \pm 0.8 \times 10^6$ M^{-1} s^{-1}	*(53)*
5′-ACCCGCGTCC[O^6-Me-G]CGCCATGGCC-3′ 3′-TGGGCGCAGG ^{Me}C GCGGTACCGG-5′	$k = 11.3 \pm 0.8 \times 10^6$ M^{-1} s^{-1}	*(53)*
5′-ACCCGCGTC^{Me}C[O^6-Me-G]CGCCATGGCC-3′ 3′-TGGGCGCAG G ^{Me}C GCGGTACCGG-5′	$k = 9.6 \pm 0.5 \times 10^6$ M^{-1} s^{-1}	*(53)*
5′-GCATGGGC[O^6-Me-G]GCATGAACCG-3′ 3′-CGTACCCG C CGTACTTGGC-5′	$k = 7.3 \pm 0.6 \times 10^6$ M^{-1} s^{-1}	*(53)*
5′-GCATGGG^{Me}C[O^6-Me-G]GCATGAACCG-3′ 3′-CGTACCC G C CGTACTTGGC-5′	$k = 12.6 \pm 1.1 \times 10^6$ M^{-1} s^{-1}	*(53)*
5′-GCATGGGC[O^6-Me-G]GCATGAACCG-3′ 3′-CGTACCCG ^{Me}C CGTACTTGGC-5′	$k = 19.2 \pm 1.0 \times 10^6$ M^{-1} s^{-1}	*(53)*
5′-GCATGGG^{Me}C[O^6-Me-G]GCATGAACCG-3′ 3′-CGTACCC G ^{Me}C CGTACTTGGC-5′	$k = 4.5 \pm 0.3 \times 10^6$ M^{-1} s^{-1}	*(53)*
5′-CATGAACC[O^6-Me-G]GAGGCCCATC-3′ 3′-GTACTTGG C CTCCGGGTAG-5′	$k = 19.7 \pm 1.0 \times 10^6$ M^{-1} s^{-1}	*(53)*
5′-CATGAAC^{Me}C[O^6-Me-G]GAGGCCCATC-3′ 3′-GTACTTG G C CTCCGGGTAG-5′	$k = 7.3 \pm 0.7 \times 10^6$ M^{-1} s^{-1}	*(53)*
5′-CATGAACC[O^6-Me-G]GAGGCCCATC-3′ 3′-GTACTTGG ^{Me}C CTCCGGGTAG-5′	$k = 14.5 \pm 1.6 \times 10^6$ M^{-1} s^{-1}	*(53)*

5′-CATGAAC^{Me}C[O^6-Me-G]GAGGCCCATC-3′ 3′-GTACTTG G ^{Me}C CTCCGGGTAG-5′	$k = 11.3 \pm 0.6 \times 10^6$ M^{-1} s^{-1}	(*53*)
5′-GTAGTTGGAGCT[O^6-Me-G]GTGGCGTAGGCAAGAGT-3′ 3′-CATCAACCTCGA C CACCGCATCCGTTCTCA-5′	$k = 1.4 \times 10^7$ M^{-1} s^{-1}	(*21*)
5′-GTAGTTGGAGCTG[O^6-Me-G]TGGCGTAGGCAAGAGT-3′ 3′-CATCAACCTCGAC C ACCGCATCCGTTCTCA-5′	$k = 7.4 \times 10^6$ M^{-1} s^{-1}	(*21*)
5′-T[O^6-Me-G]CGTGAAGTGAGTGA-3′ 3′-A C GCACTTCACTCACT-5′	ED_{50} = 7.3 nM	(*19*)
5′-TGCGT[O^6-Me-G]AAGTGAGTGA-3′ 3′-ACGCA C TTCACTCACT-5′	ED_{50} = 5.3 nM	(*19*)
5′-TGCGTGAAGT[O^6-Me-G]AGTGA-3′ 3′-ACGCACTTCA C TCACT-5′	ED_{50} = 4.2 nM	(*19*)
5′-TGCGTGAAGTG[O^6-Me-G]ATGA-3′ 3′-ACGCACTTCAC C TACT-5′	ED_{50} = 9.5 nM	(*19*)
5′-TGCGTGAAGTGA[O^6-Me-G]TGA-3′ 3′-TCGCACTTCACT C ACT-5′	ED_{50} = 7.4 nM	(*19*)
5′-TGCGTGAAGTGAG[O^6-Me-G]TA-3′ 3′-ACGCACTTCACTC C AT-5′	ED_{50} = 9.2 nM	(*19*)
5′-TGCGTGAAGTGAGT[O^6-Me-G]A-3′ 3′-ACGCACTTCACTCA C T-5′	ED_{50} = >50000 nM	(*19*)
5′-CGC[O^6-Me-G]AGCT C GCG -3′ 3′-GCG C TCGA[O^6-Me-G]CGC-5′	$k = 490 \pm 34.2 \times 10^5$ M^{-1} s^{-1}	(*41*)
5′-CG C [O^6-Me-G]CG-3′ 3′-GC[O^6-Me-G] C GC-5′	$k = 3.44 \pm 0.13 \times 10^5$ M^{-1} s^{-1}	(*41*)

Continued on next page.

Table 1. (Continued). Repair of O^6-Me-dG lesions in double stranded DNA

5′-ATGAACC[O^6-Me-G]GAGGCCCATC-3′ 3′-TACTTGG C CTCCGGGTAG-5′	$k = 3.96 \times 10^5$ M^{-1} s^{-1}	(*17*)
5′-ATGAAC^{Me}C[O^6-Me-G]GAGGCCCATC-3′ 3′-TACTTG G C CTCCGGGTAG-5′	$k = 1.18 \times 10^5$ M^{-1} s^{-1}	(*17*)
5′-TCACTTATCC[O^6-Me-G]GATACAGTA-3′ 3′-AGTGAATAGG C CTATGTCAT-5′	$k = 3.6 \times 10^7$ M^{-1} s^{-1}	(*17*)
5′-TCACTTATCC[O^6-Me-G]GATACAGTA-3′ 3′-AGTGAATAGG ^{Me}C CTATGTCAT-5′	$k = 2.1 \times 10^7$ M^{-1} s^{-1}	(*17*)
5′-TCACTTATC^{Me}C[O^6-Me-G]GATACAGTA-3′ 3′-AGTGAATAG G C CTATGTCAT-5′	$k = 0.0$ M^{-1} s^{-1}	(*17*)
5′-TCACTTATC^{Me}C[O^6-Me-G]GATACAGTA-3′ 3′-AGTGAATAG G ^{Me}C CTATGTCAT-5′	$k = 0.0$ M^{-1} s^{-1}	(*17*)
methyl[^{3}H]methylated duplex calf thymus DNA [a]	$k = 2.2 \times 10^8$ M^{-1} min^{-1}	(*18*)
5′-CGCGAATTC[O^6-Me-G]CG-3′ 3′-GCGCTTAAG C GC-5′	$k = 7.6 \times 10^8$ M^{-1} h^{-1}	(*16*)
5′-CGCCAATTG[O^6-Me-G]CG-3′ 3′-GCGGTTAAC C GC-5′	$k = 3.7 \times 10^8$ M^{-1} h^{-1}	(*16*)

[a] Prepared according to the procedures of Karran, *et al.* and Boulden, *et al.* (*61*, *62*).

Table 2. Repair of O^6-Me-dG lesions in single stranded DNA

5′-ACCCGCGTCC[O^6-Me-G]CGCCATGGCC-3′	$k = 0.7 \pm 0.05 \times 10^6\ M^{-1}\ s^{-1}$	*(53)*
5′-ACCCGCGTC^{Me}C[O^6-Me-G]CGCCATGGCC-3′	$k = 1.2 \pm 0.1 \times 10^6\ M^{-1}\ s^{-1}$	*(53)*
5′-GCATGGGC[O^6-Me-G]GCATGAACCG-3′	$k = 2.4 \pm 0.4 \times 10^6\ M^{-1}\ s^{-1}$	*(53)*
5′-GCATGGG^{Me}C[O^6-Me-G]GCATGAACCG-3′	$k = 2.4 \pm 0.2 \times 10^6\ M^{-1}\ s^{-1}$	*(53)*
5′-CATGAACC[O^6-Me-G]GAGGCCCATC-3′	$k = 0.08 \pm 0.006 \times 10^6\ M^{-1}\ s^{-1}$	*(53)*
5′-CATGAAC^{Me}C[O^6-Me-G]GAGGCCCATC-3′	$k = 0.04 \pm 0.002 \times 10^6\ M^{-1}\ s^{-1}$	*(53)*
5′-T[O^6-Me-G]CGTGAAGTGAGTGA-3′	ED_{50} = 9.5 nM	*(19)*
5′-TGCGT[O^6-Me-G]AAGTGAGTGA-3′	ED_{50} = 13.0 nM	*(19)*
5′-TGCGTGAAGT[O^6-Me-G]AGTGA-3′	ED_{50} = 11.5 nM	*(19)*
5′-TGCGTGAAGTG[O^6-Me-G]ATGA-3′	ED_{50} = 20.8 nM	*(19)*
5′-TGCGTGAAGTGA[O^6-Me-G]TGA-3′	ED_{50} = 54.5 nM	*(19)*
5′-TGCGTGAAGTGAG[O^6-Me-G]TA-3′	ED_{50} = 57.0 nM	*(19)*
5′-TGCGTGAAGTGAGT[O^6-Me-G]A-3′	ED_{50} = >50000 nM	*(19)*
5′-TmpGTGA[O^6-Me-G]CTGTmpG-3′	ED_{50} = 10 nM	*(57)*
5′-TATAC[O^6-Me-G]TATA-3′	$k = 2.00 \times 10^6\ M^{-1}\ s^{-1}$	*(15)*
5′-ATAC[O^6-Me-G]TAT-3′	$k = 2.95 \times 10^6\ M^{-1}\ s^{-1}$	*(15)*
5′-TAC[O^6-Me-G]TAT-3′	$k = 3.08 \times 10^6\ M^{-1}\ s^{-1}$	*(15)*

Continued on next page.

Table 2. (Continued). Repair of *O*⁶-Me-dG lesions in single stranded DNA

5′-TAC[O^6-Me-G]TA-3′	$k = 2.10 \times 10^5\ M^{-1}\ s^{-1}$	(*15*)
5′-AC[O^6-Me-G]TA-3′	$k = 1.90 \times 10^5\ M^{-1}\ s^{-1}$	(*15*)
5′-AC[O^6-Me-G]T-3′	$k = 2.97 \times 10^2\ M^{-1}\ s^{-1}$	(*15*)
5′-C[O^6-Me-G]T-3′	$k = 17.66\ M^{-1}\ s^{-1}$	(*15*)
5′-CGC[O^6-Me-G]CG-3′	$k = 3.44 \times 10^5\ M^{-1}\ s^{-1}$	(*15*)
5′-[O^6-Me-G]C-3′	$k = 11.36\ M^{-1}\ s^{-1}$	(*15*)
5′-[O^6-Me-G]T-3′	$k = 5.80\ M^{-1}\ s^{-1}$	(*15*)
5′-[O^6-Me-G]G-3′	$k = 4.57\ M^{-1}\ s^{-1}$	(*15*)
5′-[O^6-Me-G]A-3′	$k = 2.40\ M^{-1}\ s^{-1}$	(*15*)
5′-G[O^6-Me-G]-3′	$k = 1.08\ M^{-1}\ s^{-1}$	(*15*)
5′-C[O^6-Me-G]-3′	$k = 0.72\ M^{-1}\ s^{-1}$	(*15*)
5′-A[O^6-Me-G]-3′	$k = 0.57\ M^{-1}\ s^{-1}$	(*15*)
5′-T[O^6-Me-G]-3′	$k = 0.37\ M^{-1}\ s^{-1}$	(*15*)
5′-C[O^6-Me-G]-3′	$k = 0.71 \pm 0.02\ M^{-1}\ s^{-1}$	(*54*)
5′-T[O^6-Me-G]-3′	$k = 0.37 \pm 0.03\ M^{-1}\ s^{-1}$	(*54*)
5′-[O^6-Me-G]C-3′	$k = 11.36 \pm 0.035\ M^{-1}\ s^{-1}$	(*54*)
5′-[O^6-Me-G]T-3′	$k = 5.80 \pm 0.47\ M^{-1}\ s^{-1}$	(*54*)
methyl[^3H]methylated single-strand calf thymus DNA [a]	$k = 0.067 \times 10^8\ M^{-1}\ min^{-1}$	(*18*)

[a] Prepared according to the procedures of Karran, *et al.* and Boulden, *et al.* (*61*, *62*).

Table 3. Repair of *O^6*-Bz-dG lesions in single stranded and double stranded DNA

5′-GCCTCGAGCCAGCCGCAGACGCAG C GAGGA-3′ 3′-CGGAGCTCGGTCGGCGTCTGCGUC[O^6-Bz-G]CTCCTGCGGCT-5′	$k = 320 \times 10^6$ M^{-1} s^{-1}	(*14*)
5′-AACAGCCATAT[O^6-Bz-G]GCCC-3′	Ratio of O^6-Me-G/O^6-Bz-G repair = 0.04 ± 0.04	(*56*)
O^6-Bz-G	ED_{50} = 200 nM	(*19*)
5′-T[O^6-Bz-G][O^6-Bz-G][O^6-Bz-G][O^6-Bz-G][O^6-Bz-G]G-3′	ED_{50} = 0.7 nM	(*19*)
5′-T[O^6-Bz-G]GGGGG-3′	ED_{50} = 5.0 nM	(*19*)
5′-TGG[O^6-Bz-G]GGG-3′	ED_{50} = 3.7 nM	(*19*)
5′-TGGGG[O^6-Bz-G]G-3′	ED_{50} = 104 nM	(*19*)
5′-G[O^6-Bz-G]GGGGT-3′	ED_{50} = 3.2 nM	(*19*)
5′-GGGGG[O^6-Bz-G]T-3′	ED_{50} = 82 nM	(*19*)
O^6-Bz-G	ED_{50} = 400 nM	(*57*)
5′-TmpGTGA[O^6-Bz-G]CTGTmpG-3′	ED_{50} = 1.3 nM	(*57*)
5′-TmpGTGA[O^6-Bz-G]CTGTmpG-3′	ED_{50} = 1.4 nM	(*57*)
5′-AACAGCCATAT[O^6-Bz-G]GCCC-3′	ED_{50} = 1.1 nM	(*57*)
O^6-Bz-G	ED_{50} = 200 nM	(*43*)
O^6-Bz-dG	ED_{50} = 2000 nM	(*43*)
5′-A[O^6-Bz-G]C-3′	ED_{50} = 90 nM	(*43*)
5′-GA[O^6-Bz-G]CT-3′	ED_{50} = 13 nM	(*43*)
5′-TGA[O^6-Bz-G]CTG-3′	ED_{50} = 7 nM	(*43*)
5′-GTGA[O^6-Bz-G]CTGT-3′	ED_{50} = 8 nM	(*43*)
5′-TGTGA[O^6-Bz-G]CTGTG-3′	ED_{50} = 13 nM	(*43*)

Table 4. Repair of O^6-POB-dG and O^6-Bu-dG lesions in double stranded and single stranded DNA

5′-AACAGCCATAT[O^6-POB-G]GCCC-3′ 3′-TTGTCGGTATA C CGGG-5′	Ratio of O^6-Me-G/O^6-POB-G repair = 2 ± 0.5	(*55*)
5′-GGCGCT[O^6-POB-G]GAGGCGTG-3′ 3′-CCGCGA C CTCCGCAC-5′	Ratio of O^6-Me-G/O^6-POB-G repair = >5	(*55*)
5′-GGCGCTG[O^6-POB-G]AGGCGTG-3′ 3′-CCGCGAC C TCCGCAC-5′	Ratio of O^6-Me-G/O^6-POB-G repair = >5	(*55*)
5′-AATAGTAGCT[O^6-POB-G]GAGGC-3′ 3′-TTATCATCGA C CTCCG-5′	Ratio of O^6-Me-G/O^6-POB-G repair = 1.7 ± 0.3	(*55*)
5′-AATAGTAGCTG[O^6-POB-G]AGGC-3′ 3′-TTATCATCGAC C TCCG-5′	Ratio of O^6-Me-G/O^6-POB-G repair = >5	(*55*)
5′-AACAGCCATAT[O^6-POB-G]GCCC-3′ 3′-TTGTCGGTATA T CGGG-5′	Ratio of O^6-Me-G/O^6-POB-G repair = 2.4 ± 0.1	(*55*)
5′-AATAGTAGCTG[O^6-POB-G]AGGC-3′ 3′-TTATCATCGAC T TCCG-5′	Ratio of O^6-Me-G/O^6-POB-G repair = 1.3, 1.8	(*55*)
5′-AACAGCCATAT[O^6-POB-G]GCCC-3′	Ratio of O^6-Me-G/O^6-POB-G repair = 2.6 ± 0.5	(*56*)
5′-AACAGCCATAT[O^6-Bu-G]GCCC-3′	Ratio of O^6-Me-G/O^6-Bu-G repair = 2.4 ± 0.2	(*56*)

Table 5. Repair of O^6-Et-dG lesions in double stranded DNA

5′-ACTGACTGATGTTTGTT[O^6-Et-G]TGACTGACTG-3′ 3′-TGACTGACTACAAACAA C ACTGACTGAC-5′	$k = 10.6 \pm 4.5 \times 10^6\ M^{-1}\ s^{-1}$	(*20*)
5′-ACTGACTGATGTTTGTC[O^6-Et-G]GGACTGACTG-3′ 3′-TGACTGACTACAAACAG C CCTGACTGAC-5′	$k = 9.7 \pm 1.0 \times 10^6\ M^{-1}\ s^{-1}$	(*20*)
5′-ACTGACTGATGTTTGTG[O^6-Et-G]TGACTGACTG-3′ 3′-TGACTGACTACAAACAC C ACTGACTGAC-5′	$k = 6.5 \pm 1.3 \times 10^6\ M^{-1}\ s^{-1}$	(*20*)
5′-ACTGACTGATGTTTGTG[O^6-Et-G]GGACTGACTG-3′ 3′-TGACTGACTACAAACAC C CCTGACTGAC-5′	$k = 10.3 \pm 2.3 \times 10^6\ M^{-1}\ s^{-1}$	(*20*)
5′-ACTGACTGATGTTTGTG[O^6-Et-G]AGACTGACTG-3′ 3′-TGACTGACTACAAACAC C TCTGACTGAC-5′	$k = 11.5 \pm 5.6 \times 10^6\ M^{-1}\ s^{-1}$	(*20*)
5′-ACTGACTGATGTTTGTG[O^6-Et-G]AGACTGACTG-3′ 3′-TGACTGACTACAAACAC T TCTGACTGAC-5′	$k = 4.0 \pm 1.7 \times 10^6\ M^{-1}\ s^{-1}$	(*20*)
5′-CGC[O^6-Et-G]AGCT C GCG-3′ 3′-GCG C TCGA[O^6-Et-G]CGC-5′	$k = 3.00 \pm 0.10 \times 10^5\ M^{-1}\ s^{-1}$	(*41*)

Table 6. Repair of O^6-alkyl-dG lesions in double stranded DNA determined by first-order kinetics gave k_{inact} and K_m values

5′-GGCGCT[O^6-Me-G]GAGGCGTG-3′ 3′-CCGCGA C CTCCGCAC-5′	$k_{inact} = 0.32 \pm 0.03$ s^{-1} $K_m = 69 \pm 15$ nM	*(33)*
5′-GGCGCTG[O^6-Me-G]AGGCGTG-3′ 3′-CCGCGAC C TCCGCAC-5′	$k_{inact} = 0.027 \pm 0.009$ s^{-1} $K_m = 133 \pm 29$ nM	*(33)*
5′-AACAGCCATAT[O^6-Me-G]GCCC-3′ 3′-TTGTCGGTATA C CGGG-5′	$k_{inact} = 0.13 \pm 0.002$ s^{-1} $K_m = 112 \pm 43$ nM	*(33)*
5′-GCCTCGAGCCAGCCGCAGACGCAG C GAGGA-3′ 3′-CGGAGCTCGGTCGGCGTCTGCGUC[O^6-Me-G]CTCCTGCGGCT-5′	$k_r = 0.30 \pm 0.02$ s^{-1} $K_d = 98 \pm 32$ nM $k_r/K_d = 3.1 \times 10^6$ M^{-1} s^{-1}	*(14)*
5′-CGTGGCGCT[O^6-Me-G]GAGGCGTGAGC-3′ 3′-GCACCGCGA C CTCCGCACTCG-5′	$k_{inact} = 0.39 \pm 0.03$ s^{-1} $K_s = 81 \pm 11$ nM	*(40)*
5′-CGTGGCGCTG[O^6-Me-G]AGGCGTGAGC-3′ 3′-GCACCGCGAC C TCCGCACTCG-5′	$k_{inact} = 0.035 \pm 0.002$ s^{-1} $K_s = 91 \pm 12$ nM	*(40)*
5′-GGCGCT[O^6-POB-G]GAGGCGTG-3′ 3′-CCGCGA C CTCCGCAC-5′	$k_{inact} = (0.95 \pm 0.07) \times 10^{-4}$ s^{-1} $K_m = 125 \pm 27$ nM	*(33)*
5′-GGCGCTG[O^6-POB-G]AGGCGTG-3′ 3′-CCGCGAC C TCCGCAC-5′	$k_{inact} = (0.16 \pm 0.02) \times 10^{-4}$ s^{-1} $K_m = 161 \pm 46$ nM	*(33)*
5′-AACAGCCATAT[O^6-POB-G]GCCC-3′ 3′-TTGTCGGTATA C CGGG-5′	$k_{inact} = 0.022 \pm 0.001$ s^{-1} $K_m = 67 \pm 14$ nM	*(33)*
5′-GGCGCT[O^6-Et-G]GAGGCGTG-3′ 3′-CCGCGA C CTCCGCAC-5′	$k_{inact} = 0.023 \pm 0.001$ s^{-1} $K_m = 53 \pm 6$ nM	*(33)*

5′-GGCGCTG[O^6-Et-G]AGGCGTG-3′ 3′-CCGCGAC C TCCGCAC-5′	k_{inact} fast = 0.029 ± 0.001 s^{-1} k_{inact} slow=(2.2 ± 0.2)×10^{-4} s^{-1} K_m fast = 70 ± 46 nM K_m slow = 46 ± 11 nM	*(33)*
5′-AACAGCCATAT[O^6-Et-G]GCCC-3′ 3′-TTGTCGGTATA C CGGG-5′	k_{inact} = 0.027 ± 0.009 s^{-1} K_m = 62 ± 13 nM	*(33)*
5′-GGCGCT[O^6-Bz-G]GAGGCGTG-3′ 3′-CCGCGA C CTCCGCAC-5′	k_{inact} = 34 ± 4 s^{-1} K_m = 340 ± 90 nM[a]	*(33)*
5′-GGCGCTG[O^6-Bz-G]AGGCGTG-3′ 3′-CCGCGAC C TCCGCAC-5′	k_{inact} = 27 ± 4 s^{-1} K_m = 204 ± 87 nM[a]	*(33)*
5′-AACAGCCATAT[O^6-Bz-G]GCCC-3′ 3′-TTGTCGGTATA C CGGG-5′	k_{inact} = 52 ± 12 s^{-1} K_m = 750 ± 323 nM[a]	*(33)*
5′-GGCGCT[O^6-He-G]GAGGCGTG-3′ 3′-CCGCGA C CTCCGCAC-5′	k_{inact} = (9.3 ± 0.7) × 10^{-4} s^{-1} K_m = 210 ± 50 nM	*(33)*
5′-GGCGCTG[O^6-He-G]AGGCGTG-3′ 3′-CCGCGAC C TCCGCAC-5′	k_{inact} = (8.4 ± 0.3) × 10^{-4} s^{-1} K_m = 58 ± 11 nM	*(33)*

[a] K_m may not represent the dissociation constant since the reaction with these substrates does not reach steady state.

As discussed above, endogenous cytosine methylation plays an important role in many physiological events (*46*) and has been shown to influence the efficiency of DNA adduct formation and repair at neighboring guanine bases (*17*, *47–53*). The effect of neighboring 5–methylcytosine (^{Me}C) on the repair of O^6-Me-dG by human AGT in the context of the *p53* tumor suppressor gene was analyzed under second-order conditions (Table 1) (*17*, *53*). Bentivegna *et al.* observed a 75%-100% reduction in the rate of repair of O^6-Me-dG in a *p53*-derived sequence containing codon 248 and a designed sequence when ^{Me}C was placed 5′ to the O^6-Me-dG lesion (Table 1) (*17*). When ^{Me}C was base paired to O^6-Me-dG, the rate of repair was reduced slightly (*17*). Our laboratory has demonstrated that the kinetics of AGT-mediated repair of O^6-Me-dG is affected by neighboring ^{Me}C in a sequence-dependent manner (*53*). AGT repair of O^6-Me-dG adducts placed within 5′-CG-3′ dinucleotides in two different sequence contexts derived from the *p53* gene containing codon 245 or 248 (5′-GGC[O^6-Me-G]GC-3′ or 5′-ACC[O^6-Me-G]GA-3′) was hindered when ^{Me}C was present in both the 5′ and base paired position relative to the O^6-Me-dG lesion (Table 1). In contrast, cytosine methylation within a *p53*-derived sequence containing codon 158 (5′-TCC[O^6-Me-G]CG-3′) slightly increased the rate of O^6-Me-dG repair by AGT (Table 1) (*53*).

AGT repairs both double stranded and single stranded DNA (Tables 1 and 2). Repair of single stranded DNA may occur *in vivo* during DNA replication and may be the last opportunity to prevent promutagenic O^6-Me-G lesions from inducing mutations (*23*). However, AGT mediated repair of methylated DNA is 10-100 times more rapid in double stranded DNA than in single stranded DNA (*18*, *19*, *53*). Liem *et al.* and Wong *et al.* analyzed the effect of sequence context on AGT mediated repair of O^6-Me-dG lesions in single stranded DNA dimers under second-order kinetic conditions (Table 2) (*15*, *54*). Dimers of the sequence 5′-[O^6-Me-dG]N-3′, where N is any base (A, C, T, or G) were repaired faster than dimers with the sequence 5′-N[O^6-Me-dG]-3′ (*15*, *54*). Additionally, the rate of O^6-Me-dG repair was more sensitive to the position of neighboring pyrimidine bases than neighboring purine bases (Table 2) (*15*). The presence of ^{Me}C has also been shown to have an effect on AGT mediated repair of O^6-Me-dG lesions in single stranded DNA. AGT repair of O^6-Me-G in the context of a single stranded *p53*-derived sequence containing codon 158 is facilitated by the presence of 5′-^{Me}C, while in the context of a single stranded sequence containing codon 248 or 245 repair is inhibited or unaffected, respectively, when ^{Me}C is introduced next to O^6-Me-G (Table 2) (*53*).

An even greater effect of DNA sequence content on AGT repair efficiency was observed for O^6-POB-dG lesions induced by the tobacco specific carcinogen, NNK. The rate of AGT repair of O^6-POB-dG lesions in an *H-ras*-derived sequence containing the mutational hotspot codon 12 was investigated (Tables 4 and 6) (*33*, *55*). The kinetic parameter, k_{inact}, determined under first order conditions for duplex 15-mers containing O^6-POB-dG, was strongly influenced by sequence context (*33*). When O^6-POB-dG was placed in a run of two Gs in duplex DNA, the rate of repair was greater when the O^6-POB-dG lesion was at the 5′-guanine (Table 6) (*33*). Similarly, Mijal *et al.*, examined the ratio of O^6-Me-dG to O^6-POB-dG repair by AGT as a function of DNA sequence and found that the repair of O^6-POB-dG in an *H-ras*-like sequence occurred faster when the lesion was in the first position in a run of two Gs as opposed to the second position (Table 4) (*55*). Additionally, there was an increase in repair of *O*6-POB-dG relative to *O*6-Me-dG when T was placed opposite the *O*6-alkyl-dG adduct (*55*).

In contrast to *O*6-Me-dG and *O*6-POB-dG, the rate of AGT repair of *O*6-Et-dG and *O*6-He-dG was found to be unaffected by sequence context (Table 5 and 6). Bender *et al.* observed that AGT repair of *O*6-Et-dG in 28-mer duplexes with varying adjacent bases was not dependent on sequence (Table 5) (*20*). Although the rate of repair decreased when *O*6-Et-dG was base paired to T instead of C, this decrease was not statistically significant (*20*). Coulter *et al.* also reported that the rate of repair of *O*6-Et-dG and *O*6-He-dG by AGT was similarly unaffected by local DNA sequence (*33*). However, when *O*6-Et-dG was the second G in a run of two Gs the time course of repair was biphasic (Table 6) (*33*). This biphasic kinetics was attributed to binding of *O*6-Et-dG to AGT in a reactive or a less reactive conformation and was found to be affected by neighboring nucleotides (*33*).

Similar to *O*6-Et-dG and *O*6-He-dG lesions, the sequence context of *O*6-Bz-dG lesions had little effect on their rates of repair by AGT (Table 3). However, *O*6-Bz-dG lesions, both in single stranded and in double stranded DNA, were repaired more rapidly than other *O*6-alkyl-dG lesions (*14*, *56*). Luu *et al.* and Coulter *et al.* reported that the repair of *O*6-Bz-dG by human AGT was independent of the bases flanking the *O*6-Bz-dG lesion (*19*, *33*). However, the position of the *O*6-Bz-dG lesion and the number of bases flanking the lesion in single stranded DNA affects the efficiency of AGT mediated repair (*19*, *43*). The rate of AGT mediated repair of *O*6-Bz-dG in the central position of a single stranded oligonucleotide increased as the length of the oligonucleotide increased, with the maximal rate achieved when the oligomer was 5-9 nucleotides long (*43*).

The relative rates of AGT-mediated repair of *O*6-alkyl-dG lesions in duplex DNA are benzyl > methyl > ethyl >> 2-hydroxyethyl >4-(3-pyridyl)-4-oxobutyl (*33*). AGT mediated repair of *O*6-Et-dG is 10-100-fold less efficient than repair of *O*6-Me-dG ((*33*, *41*). Zang *et al.* demonstrated that in duplex DNA, *O*6-Bz-dG repair by AGT is about 100-fold faster than *O*6-Me-dG repair (*14*). In single stranded DNA, Mijal el al. compared the ratio of *O*6-Me-dG repair to repair of other *O*6-alkyl-dG lesions and found that AGT preferentially repairs *O*6-Me-dG lesions relative to *O*6-POB-dG and *O*6-Bu-dG (*56*). In contrast, *O*6-Bz-dG is preferentially repaired relative to *O*6-Me-dG (*19*, *56*, *57*).

Effect of DNA Sequence on AGT Protein Affinity

The first step of AGT mediated repair that may be influenced by sequence content is binding of the AGT protein to DNA (Step 1 in Figure 2). The rate of AGT binding to DNA was very fast and appeared to be diffusion-limited (5×10^9 M^{-1} s^{-1}) (*14*), while the rate of dissociation is slower (*14*). The presence of an *O*6-alkyl-dG lesion slows the dissociation about 25-fold slower as compared to unmodified DNA (*14*).

Table 7. Binding affinity of AGT for single stranded DNA lacking and containing O^6-alkyl-dG lesions

5′-TTTTTGTTTTT-3′	$K_n = 3.20 \times 10^{10}$ M^{-n} Stoichiometry = 1.92 ± 0.17 $K_{mono} = 3.33 \pm 1.79 \times 10^5$ M^{-1}	(*58*)
5′-GACTGACTGACTGACT-3′	$K_n = 9.42 \times 10^{22}$ M^{-n} Stoichiometry = 3.89 ± 0.09 $K_{mono} = 8.05 \pm 0.26 \times 10^5$ M^{-1}	(*58*)
5′-CGCCAACCCGCTGCCTATCGTT-3′	$K_n = 1.55 \times 10^{22}$ M^{-n} Stoichiometry = 3.89 ± 0.16 $K_{mono} = 4.42 \pm 0.88 \times 10^5$ M^{-1}	(*58*)
5′-TTTTTTTTTTTTTTTTTTTTTTTT-3′	$K_n = 9.64 \times 10^{31}$ M^{-n} Stoichiometry = 5.60 ± 0.5 $K_{mono} = 5.35 \pm 1.98 \times 10^5$ M^{-1}	(*58*)
5′-GTGCCGCCAACCCGCTGCCTATCGTTATAC-3′	$K_n = 1.03 \times 10^{30}$ M^{-n} Stoichiometry = 5.32 ± 0.42 $K_{mono} = 4.53 \pm 1.55 \times 10^5$ M^{-1}	(*58*)
5′-GCAACGCAATTAATGTGAGTTAGCTCACTCATTAGCCACCC-3′	$K_n = 1.34 \times 10^{33}$ M^{-n} Stoichiometry = 5.89 ± 1.1 $K_{mono} = 4.14 \pm 0.69 \times 10^5$ M^{-1}	(*58*)
5′-GACTGACTGACTGACT-3′	$K_{PD} = 1.8 (\pm 0.7) \times 10^{24}$ M^{-4} Stoichiometry = 3.8 ± 0.3 $K_{mono} = \sim 1.2 \times 10^6$ M^{-1} (wt-AGT)	(*23*)
5′-GTGCCGCCAACCCGCTGCCTATCGTTATAC-3′	Stoichiometry = 5.3 ± 0.2 (wt-AGT)	(*23*)

5′-GTGGTCTGCAGCAGCGGAGCCGTGGGCAACTACCGCTGGGGCGTGTC GCGTAAGGAATGGCTTCTGGCCCATGAAGGC-3′	Stoichiometry = 8.9 ± 0.2 (wt-AGT)	(*23*)
5′-GACTGACTGACTGACT-3′	K_a = ~1.2 × 10^{23} M^{-4} Stoichiometry = 4 K_{mono} = ~5.9 × 10^5 M^{-1} (C145A-AGT)	(*22*)
5′-GACTGACTGACT[O^6-Me-G]ACT-3′	K_a = ~21 × 10^{24} M^{-4} Stoichiometry = 4 K_{mono} = ~21 × 10^5 M^{-1} (C145A-AGT)	(*22*)
5′-AACAGCCATATGGCCC-3′	K_d = 0.90 μM (C145A-AGT) K_d = 1.1 μM (wt-AGT)	(*60*)
5′-AACAGCCATAT[O^6-Me-G]GCCC-3′	K_d = 0.70 μM (C145A-AGT)	(*60*)
5′-ACTGACTGATGTTTGTCGGGACTGACTG-3′	K_a = 1.0 ± 0.4 × 10^5 M^{-1}	(*20*)
5′-ACTGACTGATGTTTGTC[O^6-Et-G]GGACTGACTG-3′	K_a = 4.9 ± 2.0 × 10^5 M^{-1}	(*20*)
5′-TGGGGGG-5′	K_d = 3.9 μM (C145S-AGT)	(*19*)
5′-T[O^6-Bz-G][O^6-Bz-G][O^6-Bz-G][O^6-Bz-G][O^6-Bz-G]G-3′	K_d = 3.7 μM (C145S-AGT)	(*19*)
5′-T[O^6-Bz-G]GGGGG-3′	K_d = 2.9 μM (C145S-AGT)	(*19*)
5′-TGG[O^6-Bz-G]GGG-3′	K_d = 25 μM (C145S-AGT)	(*19*)
5′-TGGGG[O^6-Bz-G]G-3′	K_d = 4.4 μM (C145S-AGT)	(*19*)

Table 8. Binding affinity of AGT for double stranded DNA lacking and containing *O*^{6}-alkyl-dG lesions

5′-ACCCGCGTCC[O^6-Me-G]CGCCATGGCC-3′ 3′-TGGGCGCAGG C GCGGTACCGG-5′	K_d = 1.3 ± 0.09 × 10^{-6} M (C145A-AGT)	*(53)*
5′-ACCCGCGTC^{Me}C[O^6-Me-G]CGCCATGGCC-3′ 3′-TGGGCGCAG G C GCGGTACCGG-5′	K_d = 1.2 ± 0.04 × 10^{-6} M (C145A-AGT)	*(53)*
5′-ACCCGCGTCC[O^6-Me-G]CGCCATGGCC-3′ 3′-TGGGCGCAGG ^{Me}C GCGGTACCGG-5′	K_d = 1.1 ± 0.09 × 10^{-6} M (C145A-AGT)	*(53)*
5′-ACCCGCGTC^{Me}C[O^6-Me-G]CGCCATGGCC-3′ 3′-TGGGCGCAG G ^{Me}C GCGGTACCGG-5′	K_d = 1.2 ± 0.04 × 10^{-6} M (C145A-AGT)	*(53)*
5′-GCATGGGC[O^6-Me-G]GCATGAACCG-3′ 3′-CGTACCCG C CGTACTTGGC-5′	K_d = 0.9 ± 0.06 × 10^{-6} M (C145A-AGT)	*(53)*
5′-GCATGGG^{Me}C[O^6-Me-G]GCATGAACCG-3′ 3′-CGTACCC G ^{Me}C CGTACTTGGC-5′	K_d = 0.9 ± 0.08 × 10^{-6} M (C145A-AGT)	*(53)*
5′-CATGAACC[O^6-Me-G]GAGGCCCATC-3′ 3′-GTACTTGG C CTCCGGGTAG-5′	K_d = 0.6 ± 0.06 × 10^{-6} M (C145A-AGT)	*(53)*
5′-CATGAAC^{Me}C[O^6-Me-G]GAGGCCCATC-3′ 3′-GTACTTG G ^{Me}C CTCCGGGTAG-5′	K_d = 0.8 ± 0.05 × 10^{-6} M (C145A-AGT)	*(53)*
5′-CGTGGCGCTGGAGGCGTGAGC-3′ 3′-GCACCGCGACCTCCGCACTCG-5′	760 ± 80 (C145S-AGT)	*(40)*
5′-CGTGGCGCT[O^6-Me-G]GAGGCGTGAGC-3′ 3′-GCACCGCGA C CTCCGCACTCG-5′	K_d = 380 ± 50 nM (C145S-AGT)	*(40)*
5′-CGTGGCGCTG[O^6-Me-G]AGGCGTGAGC-3′ 3′-GCACCGCGAC C TCCGCACTCG-5′	K_d = 240 ± 40 nM (C145S-AGT)	*(40)*

5′-GACTGACTGACTGACT-3′ 3′-CTGACTGACTGACTGA-5′	K_a = ~4.3 × 10^{23} M^{-4} Stoichiometry = 4 K_{mono} = ~8.0 × 10^5 M^{-1} (C145A-AGT)	*(22)*
5′-GACTGACTGACT[O^6-Me-G]ACT-3′ 3′-CTGACTGACTGA C TGA-5′	K_a = ~31 × 10^{24} M^{-4} Stoichiometry = 4 K_{mono} = ~23 × 10^5 M^{-1} (C145A-AGT)	*(22)*
5′-AACAGCCATATGGCCC-3′ 3′-TTGTCGGTATACCGGG-5′	K_d = 0.69 μM (C145A-AGT) K_d = 0.61 μM (wt-AGT)	*(60)*
5′-AACAGCCATAT[O^6-Me-G]GCCC-3′ 3′-TTGTCGGTATA C CGGG-5′	K_d = 0.13 μM (C145A-AGT)	*(60)*
calf-thymus DNA	K_a = 4.7 × 10^5 M^{-1}	*(59)*
calf-thymus DNA reacted with [^{3}H-methyl]MNU	K_a = 9 × 10^5 M^{-1}	*(59)*
5′-ACTGACTGATGTTTGTCGGGACTGACTG-3′ 3′-TGACTGACTACAAACAGCCCTGACTGAC-5′	K_a = 7.1 ± 2.4 × 10^5 M^{-1}	*(20)*
5′-ACTGACTGATGTTTGTC[O^6-Et-G]GGACTGACTG-3′ 3′-TGACTGACTACAAACAG C CCTGACTGAC-5′	K_a = 8.4 ± 2.4 × 10^5 M^{-1}	*(20)*

Table 9. Rate of AGT-induced *O*⁶-alkyl-dG nucleotide flipping

5′-ACCCGCGTCC [O^6-Me-G]CGCCATGGCC-3′ 3′-TGGGCGCAGG[pyrrolo-C]GCGGTACCGG-5′	$k = 109 \pm 21\ s^{-1}$ (C145A-AGT)	(*53*)
5′-ACCCGCGTC^{Me}C [O^6-Me-G]CGCCATGGCC-3′ 3′-TGGGCGCAG G[pyrrolo-C]GCGGTACCGG-5′	$k = 94 \pm 28\ s^{-1}$ (C145A-AGT)	(*53*)
5′-GCATGGGC [O^6-Me-G]GCATGAACCG-3′ 3′-CGTACCCG[pyrrolo-C]CGTACTTGGC-5′	$k = 182 \pm 32\ s^{-1}$ (C145A-AGT)	(*53*)
5′-GCATGGG^{Me}C [O^6-Me-G]GCATGAACCG-3′ 3′-CGTACCC G[pyrrolo-C]CGTACTTGGC-5′	$k = 162 \pm 46\ s^{-1}$ (C145A-AGT)	(*53*)
5′-CATGAACC [O^6-Me-G]GAGGCCCATC-3′ 3′-GTACTTGG[pyrrolo-C]CTCCGGGTAG-5′	$k = 80 \pm 30\ s^{-1}$ (C145A-AGT)	(*53*)
5′-CATGAAC^{Me}C [O^6-Me-G]GAGGCCCATC-3′ 3′-GTACTTG G[pyrrolo-C]CTCCGGGTAG-5′	$k = 107 \pm 35\ s^{-1}$ (C145A-AGT)	(*53*)
5′-GCCTCGAGCCAGCCGCAGACGCAG[pyrrolo-C]GAGGA-3′ 3′-CGGAGCTCGGTCGGCGTCTGCGUC [O^6-Bz-G]CTCCTGCGGCT-5′	$k = 190 \pm 20\ s^{-1}$ (C145S-AGT)	(*14*)
5′-GCCTCGAGCCAGCCGCAGACGCAG[pyrrolo-C]GAGGA-3′ 3′-CGGAGCTCGGTCGGCGTCTGCGUC [O^6-Me-G]CTCCTGCGGCT-5′	$k = 200 \pm 22\ s^{-1}$ (C145S-AGT)	(*14*)

The majority of studies that investigated AGT binding to DNA have determined association or dissociation constants using ^{32}P-end-labeled oligonucleotides or kinetic conditions that allowed for the determination of dissociation constants. Binding studies showed that AGT has affinity for single stranded and double stranded DNA with or without O^6-alkyl-dG lesions (Tables 7 and 8). Based on crystal structures, the binding site size of AGT on double stranded DNA was predicted to be 7 nucleotides long (*26*, *27*). However, binding studies completed by Fried *et al.* and Rasimas *et al.* using single stranded and double stranded DNA observed an AGT binding site size between 4 and 9 nucleotides, suggesting that several AGT molecules overlap along the DNA duplex (*22*, *23*, *58*). These findings also revealed that AGT binds cooperatively to single stranded and double stranded DNA with or without O^6-alkyl-dG lesions (*22*, *23*, *58*). Cooperative binding of AGT to DNA may be an efficient mechanism by which O^6-alkyl-dG lesions are located and repaired (*22*, *23*, *58*). In unmodified single stranded and double stranded DNA, the binding affinity of AGT was not significantly affected by DNA sequence (Table 7 and 8) (*23*, *58*).

Several researchers examined whether the presence of an O^6-alkyl-dG lesion increases the binding affinity of the AGT protein for DNA. The AGT binding affinity was enhanced 2-5 fold when the O^6-Me-dG lesion was present in single stranded or double stranded DNA (Table 7 and 8) as compared to unmethylated dG (*22*, *40*, *59*, *60*). In contrast, the binding affinity of AGT was not affected by the presence or absence of an O^6-Et-dG lesion in single stranded or double stranded DNA (*20*). The presence of an O^6-Bz-dG lesion also had no affect on the binding affinity of AGT to single stranded DNA (*19*). The most extensive comparison of AGT binding to DNA duplexes containing O^6-alkyl-dG lesions was carried out by Coulter *et al.* who found that the AGT binding affinity was not affected by the identity of the alkyl group on the O^6-alkyl-dG lesion or the nucleotide sequence surrounding the lesion (Table 6) (*33*). Taken together, binding affinity experiments suggest that AGT preferentially binds double stranded DNA. While the presence of certain O^6-alkyl-dG lesions increases the binding affinity, it appears to be generally independent of DNA sequence.

Sequence Effects on the Rate of AGT-Mediated Nucleotide Flipping

Following AGT binding to DNA, the adducted nucleotide is flipped into the active site of the protein (Step 2 of Figure 2). Zang *et al.* developed a method to determine the rate of nucleotide flipping by AGT using pre-steady state stopped-flow fluorescence experiments (*14*). In duplex DNA, the O^6-alkyl-dG lesion was placed opposite a fluorescent nucleoside analog, 6-methylpyrrolo[2,3-d]pyrimidine-2(3H)one deoxyribonucleoside (pyrrolo dC). The fluorescence intensity of the nucleoside analog increases when the O^6-alkyl-dG lesion is flipped out of the dupex by AGT, disrupting its hydrogen bonding to pyrrolo-dC. Measurement of the rate of nucleotide flipping of O^6-Me-dG and O^6-Bz-dG lesions by AGT using this technique indicated that both lesions were flipped into the active site of AGT at a similar rate (Table

9) (*14*). This method was then used by our laboratory to examine the affect of sequence context on the rate of nucleotide flipping in the presence or absence of a neighboring ^{Me}C residue i. The presence of ^{Me}C immediately 5′ to the O^6-Me-dG residue in duplex DNA did not significantly influence the rate of AGT-mediated nucleotide flipping (Table 9) (*53*).

Effects of DNA Sequence on the Rate of Alkyl Transfer from O^6-Alkyl-dG to AGT

Once the O^6-alkyl-dG lesion is flipped out of the base pair stack to enter the active site of the protein, the O^6-alkyl group is positioned for removal via an S_N2 reaction by the thiol of the active site cysteine (Figure 4). Since most studies have observed no sequence effect on AGT protein binding and nucleotide flipping, it is likely that sequence-dependent changes in the rate of alkyl transfer are responsible for the observed overall differences in dealkylation rates (*19*, *33*, *40*, *53*). Coulter *et al.* proposed that the local sequence content affects the position of the O^6-alkyl-dG lesion in the AGT binding pocket (*33*). Crystal structures have shown that the adducted nucleotide is positioned in the AGT active site for in-line displacement of the O^6-substituent by the active site cysteine (*26*, *27*). The local sequence context may result in small differences in the distance between the alkyl group and the active site cysteine, potentially affecting the rate of alkyl transfer (*33*). Experimentally the direct determination of the rate of alkyl transfer is difficult; however, Zang *et al.* were able to calculate the rate of methyl transfer after determining the rate for all of the other steps in the reaction (O^6-Me-dG, $k_{methyl\ transfer} = 25\ s^{-1}$) (*14*). It was determined that alkyl transfer is rate limiting for repair of O^6-Me-dG, but not for O^6-Bz-dG (*14*). Further work must be done to determine the mechanism of sequence dependent effect on the rate of alkyl transfer.

Conclusions

Inefficient repair of O^6-alkyl-dG lesions by AGT can lead to the accumulation of mutations at specific sites within the genome. If these mutations occur within critical genes, such as the *ras* proto-oncogenes and the *p53* tumor suppressor gene, a loss of control over cellular growth and proliferation may result, leading to cancer initiation. AGT is able to remove the alkyl group from O^6-alkyl-dG lesions in DNA, restoring normal guanine and preventing mutagenesis. The rates of AGT-mediated repair of O^6-alkyl-dG lesions are influenced by alkyl group identity, sequence length, adduct placement, and neighboring nucleotides. As a consequence, certain sites within the genome may be inefficiently repaired, leading to O^6-alkyl-dG adduct persistence and the accumulation of mutations. Studies to date have shown that binding of AGT protein to DNA and nucleotide flipping are unaffected by the DNA sequence and the alkyl group identity. This suggests that at least for some lesions, such as O^6-Me-dG, the local DNA sequence influences the rate of repair by mediating the rate of alkyl transfer.

List of Abbreviations

AGT, O^6-alkylguanine DNA alkyltransferase; HTH, helix-turn-helix; ^{Me}C, 5-methylcytosine; O^6-alkyl-dG, O^6-alkyl-deoxyguanosine; O^6-Bu-dG, O^6-butyl-deoxyguanosine; O^6-Bz-dG, O^6-benzyl-deoxyguanosine; O^6-Et-dG, O^6-ethyl-deoxyguanosine; O^6-He-dG, O^6-2-hydroxyethyl-deoxyguanosine; O^6-Me-dG, O^6-methyl-deoxyguanosine; O^6-POB-dG, O^6-4-(3-pyridyl)-4-oxobutyl-deoxyguanosine; pyrrolo-dC, 6-methylpyrrolo[2,3-d]pyrimidine-2(3H)one deoxyribonucleosid.

References

1. Loveless, A. *Nature* **1969**, *223*, 206–207.
2. Gerson, S. L. *J. Clin. Oncol.* **2002**, *20*, 2388–2399.
3. Kyrtopoulos, S. A.; Anderson, L. M.; Chhabra, S. K.; Souliotis, V. L.; Pletsa, V.; Valavanis, C; Georgiadis, P. *Cancer Detect. Prev.* **1997**, *21*, 391–405.
4. Snow, E. T.; Mitra, S. *Cancer Invest.* **1987**, *5*, 119–125.
5. Warren, J. J.; Forsberg, L. J.; Beese, L. S. *Proc. Natl. Acad. Sci. U.S.A.* **2006**, *103*, 19701–19706.
6. Westra, W. H.; Slebos, R. J.; Offerhaus, G. J.; Goodman, S. N.; Evers, S. G.; Kensler, T. W.; Askin, F. B.; Rodenhuis, S.; Hruban, R. H. *Cancer* **1993**, *72*, 432–438.
7. Ronai, Z. A.; Gradia, S.; Peterson, L. A.; Hecht, S. S. *Carcinogenesis* **1993**, *14*, 2419–2422.
8. Hussain, S. P.; Harris, C. C. *Mutat. Res.* **1999**, *428*, 23–32.
9. Wolf, P.; Hu, Y. C.; Doffek, K.; Sidransky, D.; Ahrendt, S. A. *Cancer Res.* **2001**, *61*, 8113–8117.
10. Ahrendt, S. A.; Chow, J. T.; Yang, S. C.; Wu, L.; Zhang, M. J.; Jen, J.; Sidransky, D. *Cancer Res.* **2000**, *60*, 3155–3159.
11. Esteller, M.; Toyota, M.; Sanchez-Cespedes, M.; Capella, G.; Peinado, M. A.; Watkins, D. N.; Issa, J. P.; Sidransky, D.; Baylin, S. B.; Herman, J. G. *Cancer Res.* **2000**, *60*, 2368–2371.
12. Bello, M. J.; Alonso, M. E.; Amioso, C.; Anselmo, N. P.; Arjona, D.; Gonzalez-Gomez, P.; Lopez-Marin, I.; de Campos, J. M.; Gutierrez, M.; Isla, A.; et al. *Mutat. Res.* **2004**, *554*, 23–32.
13. Nakamura, M.; Watanabe, T.; Yonekawa, Y.; Kleihues, P.; Ohgaki, H. *Carcinogenesis* **2001**, *22*, 1715–1719.
14. Zang, H.; Fang, Q.; Pegg, A. E.; Guengerich, F. P. *J. Biol. Chem.* **2005**, *280*, 30873–30881.
15. Liem, L. K.; Wong, C. W.; Lim, A.; Li, B. F. L. *J. Mol. Biol.* **1993**, *231*, 950–959.
16. Dolan, M. E.; Oplinger, M.; Pegg, A. E. *Carcinogenesis* **1988**, *9*, 2139–2143.
17. Bentivegna, S. S.; Bresnick, E. *Cancer Res.* **1994**, *54*, 327–329.
18. Bhattacharyya, D.; Foote, R. S.; Boulden, A. M.; Mitra, S. *Eur. J. Biochem.* **1990**, *193*, 337–343.

19. Luu, K. X.; Kanugula, S.; Pegg, A. E.; Pauly, G. T.; Moschel, R. C. *Biochemistry* **2002**, *41*, 8689–8697.
20. Bender, K.; Federwisch, M.; Loggen, U.; Nehls, P.; Rajewsky, M. F. *Nucleic Acids Res.* **1996**, *24*, 2087–2094.
21. Guza, R.; Rajesh, M.; Fang, Q.; Pegg, A. E.; Tretyakova, N. *Chem. Res. Toxicol.* **2006**, *19*, 531–538.
22. Rasimas, J. J.; Pegg, A. E.; Fried, M. G. *J. Biol. Chem.* **2003**, *278*, 7973–7980.
23. Fried, M. G.; Kanugula, S.; Bromberg, J. L.; Pegg, A. E. *Biochemistry* **1996**, *35*, 15295–15301.
24. Wibley, J. E. A.; Pegg, A. E.; Moody, P. C. E. *Nucleic Acids Res.* **2000**, *28*, 393–401.
25. Daniels, D. S.; Mol, C. D.; Arvai, A. S.; Kanugula, S.; Pegg, A. E.; Tainer, J. A. *EMBO J.* **2000**, *19*, 1719–1730.
26. Daniels, D. S.; Woo, T. T.; Luu, K. X.; Noll, D. M.; Clarke, N. D.; Pegg, A. E.; Tainer, J. A. *Nat. Struct. Mol. Biol.* **2004**, *11*, 714–720.
27. Duguid, E. M.; Rice, P. A.; He, C. *J. Mol. Biol.* **2005**, *350*, 657–666.
28. Guengerich, F. P.; Fang, Q.; Liu, L.; Hachey, D. L.; Pegg, A. E. *Biochemistry* **2003**, *42*, 10965–10970.
29. Rasimas, J. J.; Kanugula, S.; Dalessio, P. M.; Ropson, I. J.; Fried, M. G.; Pegg, A. E. *Biochemistry* **2003**, *42*, 980–990.
30. Pegg, A. E.; Dolan, M. E.; Moschel, R. C. *Prog. Nucleic Acid Res. Mol. Biol.* **1995**, *51*, 167–223.
31. Spratt, T. E.; Wu, J. D.; Levy, D. E.; Kanugula, S.; Pegg, A. E. *Biochemistry* **1999**, *38*, 6801–6806.
32. Georgieva, P.; Himo, F. *Chem. Phys. Lett.* **2008**, *463*, 214–218.
33. Coulter, R.; Blandino, M.; Tomlinson, J. M.; Pauly, G. T.; Krajewska, M.; Moschel, R. C.; Peterson, L. A.; Pegg, A. E.; Spratt, T. E. *Chem. Res. Toxicol.* **2007**, *20*, 1966–1971.
34. Leonard, G. A.; Thomson, J.; Watson, W. P.; Brown, T. *Proc. Natl. Acad. Sci. U.S.A.* **1990**, *87*, 9573–9576.
35. Kalnik, M. W.; Li, B. F. L.; Swann, P. F.; Patel, D. J. *Biochemistry* **1989**, *28*, 6182–6192.
36. Srivenugopal, K. S.; Yuan, X. H.; Friedman, H. S.; Ali-Osman, F. *Biochemistry* **1996**, *35*, 1328–1334.
37. Major, G. N.; Brady, M.; Notarianni, G. B.; Collier, J. D.; Douglas, M. S. *Biochem. Soc. Trans.* **1997**, *25*, 359S.
38. Xu-Welliver, M.; Pegg, A. E. *Carcinogenesis* **2002**, *23*, 823–830.
39. Tubbs, J. L.; Pegg, A. E.; Tainer, J. A. *DNA Repair* **2007**, *6*, 1100–1115.
40. Meyer, A. S.; McCain, M. D.; Fang, Q.; Pegg, A. E.; Spratt, T. E. *Chem. Res. Toxicol.* **2003**, *16*, 1405–1409.
41. Liem, L. K.; Lim, A.; Li, B. F. L. *Nucleic Acids Res.* **1994**, *22*, 1613–1619.
42. Scicchitano, D.; Jones, R. A.; Kuzmich, S.; Gaffney, B.; Lasko, D. D.; Essigmann, J. M.; Pegg, A. E. *Carcinogenesis* **1986**, *7*, 1383–1386.
43. Pegg, A. E.; Kanugula, S.; Edara, S.; Pauly, G. T.; Moschel, R. C.; Goodtzova, K. *J. Biol. Chem.* **1998**, *273*, 10863–10867.

44. Ali, R. B.; Teo, A. K. C.; Oh, H. K.; Chuang, L. S. H.; Ayi, T. C.; Li, B. F. *Mol. Cell. Biol.* **1998**, *18*, 1660–1669.
45. Begley, T. J.; Samson, L. D. *Nat. Struct. Mol. Biol.* **2004**, *11*, 688–690.
46. Riggs, A. D.; Jones, P. A. *Adv. Cancer Res.* **1983**, *40*, 1–30.
47. Denissenko, M. F.; Chen, J. X.; Tang, M. S.; Pfeifer, G. P. *Proc. Natl. Acad. Sci. U.S.A.* **1997**, *94*, 3893–3898.
48. Tretyakova, N.; Matter, B.; Jones, R.; Shallop, A. *Biochemistry* **2002**, *41*, 9535–9544.
49. Matter, B.; Wang, G.; Jones, R.; Tretyakova, N. *Chem. Res. Toxicol.* **2004**, *17*, 731–741.
50. Ziegel, R.; Shallop, A.; Jones, R.; Tretyakova, N. *Chem. Res. Toxicol.* **2003**, *16*, 541–550.
51. Ziegel, R.; Shallop, A.; Upadhyaya, P.; Jones, R.; Tretyakova, N. *Biochemistry* **2004**, *43*, 540–549.
52. Rajesh, M.; Wang, G.; Jones, R.; Tretyakova, N. *Biochemistry* **2005**, *44*, 2197–2207.
53. Guza, R.; Ma, L.; Fang, Q.; Pegg, A. E.; Tretyakova, N. *J. Biol. Chem.* **2009**, *284*, 22601–22610.
54. Wong, C. W.; Tan, N. W.; Li, B. F. L. *J. Mol. Biol.* **1992**, *228*, 1137–1146.
55. Mijal, R. S.; Kanugula, S.; Vu, C. C.; Fang, Q.; Pegg, A. E.; Peterson, L. A. *Cancer Res.* **2006**, *66*, 4968–4974.
56. Mijal, R. S.; Thomson, N. M.; Fleischer, N. L.; Pauly, G. T.; Moschel, R. C.; Kanugula, S.; Fang, Q.; Pegg, A. E.; Peterson, L. A. *Chem. Res. Toxicol.* **2004**, *17*, 424–434.
57. Pegg, A. E.; Goodtzova, K.; Loktionova, N. A.; Kanugula, S.; Pauly, G. T.; Moschel, R. C. *J. Pharmacol. Exp. Ther.* **2001**, *296*, 958–965.
58. Rasimas, J. J.; Kar, S. R.; Pegg, A. E.; Fried, M. G. *J. Biol. Chem.* **2007**, *282*, 3357–3366.
59. Chan, C. L.; Wu, Z. N.; Ciardelli, T.; Eastman, A.; Bresnick, E. *Arch. Biochem. Biophys.* **1993**, *300*, 193–200.
60. Goodtzova, K.; Kanugula, S.; Edara, S.; Pegg, A. E. *Biochemistry* **1998**, *37*, 12489–12495.
61. Karran, P.; Lindahl, T.; Griffin, B. *Nature* **1979**, *280*, 76–77.
62. Boulden, A. M.; Foote, R. S.; Fleming, G. S.; Mitra, S. *J. Biosci.* **1987**, *11*, 215–224.

Chapter 7

Impact of Carcinogen-DNA Adducts on DNA Methylation

Elizaveta S. Gromova,[1] Oksana M. Subach,[1] Vladimir B. Baskunov,[1] and Nicholas E. Geacintov[2,*]

[1]Department of Chemistry, Lomonossov Moscow State University, Moscow, 119991, Russia
[2]Department of Chemistry, New York University, 31 Washington Place, New York, New York 10003-5180, USA
[*]nicholas.geacintov@nyu.edu

The methylation of DNA at CpG dinucleotides by DNA methyltransferases (MTases) is an epigenetic alteration of the genome that plays an important role in the regulation of gene expression in eukaryotes. Abnormalities in the levels of methylation are one of the hallmarks of tumorigenesis. Polycyclic aromatic hydrocarbons such as benzo[*a*]pyrene are ubiquitous environmental pollutants that are metabolized in vivo to highly genotoxic and tumorigenic diol epoxides (B[*a*]PDE). The latter reacts with cellular DNA to form covalent adducts that can, if not removed by cellular DNA repair mechanisms, cause mutations and the initiation of tumorigenesis in animal model systems, and probably in humans. The potential impact of such lesions on DNA methylation has received relatively little attention. Utilizing an in vitro biochemical approach, it is shown that BPDE-derived DNA lesions can alter DNA methylation by prokaryotic MTases M.SssI and M.HhaI and mammalian MTase Dnmt3a at CpG and GCGC sites. These effects depend on the conformational properties of the lesions and their positions within the DNA recognition sequence. The results of these studies suggest that B[*a*]PDE may initiate cancer not only by genotoxic mechanisms, but might also contribute to tumor development by epigenetic effects that involve changes in DNA methylation status.

Benzo[*a*]pyrene (B[*a*]P) is one of the most common polycyclic aromatic hydrocarbons (PAHs) (*1*). PAHs are carcinogens that are pervasive in the human environment and result from the incomplete combustion of virtually any organic material. B[*a*]P can be metabolized into different 7,8-diol-9,10-epoxide stereoisomers, the major metabolite being the highly genotoxic (+)-7*R*,8*S*,9*S*,10*R* enantiomer of 7*r*,8*t*-dihydroxy-*t*9,10-epoxy-7,8,9,10-tetrahydrobenzo[*a*]pyrene (B[*a*]PDE), while minor amounts of the (-)-7*S*,8*R*,9*R*,10*S* enantiomer also formed (*2*). These B[*a*]PDE metabolites are highly reactive and bind predominantly and covalently with the exocyclic amino group of guanine in DNA by *trans* addition to the C10 position of B[*a*]PDE, or by *cis* addition (*3*). The most extensively studied adducts are (+)-*trans* -B[*a*]P-N^2-dG, (+)-*cis*-B[*a*]P-N^2-dG, (-)-*trans*-B[*a*]P-N^2-dG, and (-)-*cis*- B[*a*]P-N^2-dG. The structures and stereochemical properties of these and related PAH-DNA adducts have been sumnmarized (*4*). The conformational properties of three of these adducts discussed in this work, are depicted in Figure 1. The relative yield of the (+)-*trans*-B[*a*]P-N^2-dG adduct is generally the highest when racemic B[*a*]PDE is reacted with native DNA (about 90%) (*3*). In the case of the (+)-*trans* and (-)-*trans*-B[*a*]P-N^2-dG adducts, the bulky pyrenyl residues are positioned in the minor groove of intact B-DNA duplexes, with the B[*a*]P residues pointing either into the 5′- or the 3′-directions of the modified DNA strand, respectively, relative to the modified guanine residue of the B[*a*]P-N^2-dG adduct (*5*, *6*) (Figure 1). In contrast, the (+)-*cis*-B[*a*]P-N^2-dG adduct has a base-displaced intercalative conformation with the modified guanine and the partner cytosine displaced into the minor and major grooves, respectively (*7*). If not repaired, the resulting B[*a*]P-N^2-dG adducts can induce mutations during DNA replication (*8–11*). Such DNA lesions are formed efficiently at the guanine residue in CpG sequences (*12*) that are recognition sites of mammalian MTases. The efficiency of such damage is enhanced in the presence of m^{5}dC instead of dC in 5′-CpG targets (*12–14*). Therefore, the B[*a*]P-N^2-dG adducts can, in principle, also exert epigenetic effects by altering the methylation status of the modified CpG sequences and affect heritable gene expression. While the susceptibilities of different PAH diol epoxide-DNA adducts to removal by nucleotide excision repair mechanisms have been examined (*15*, *16*), and their mutagenic properties have been investigated (*12–14*), the potential impact of these lesions on DNA methylation has received little attention. Nevertheless, it has been previously shown that the fraction of methylated cytosines decreases in the DNA of mammalian cells treated with racemic B[*a*]PDE (*17*, *18*).

The prokaryotic MTases SssI (M.SssI) and HhaI (M.HhaI) transfer a methyl group to the carbon C5 of the cytosine residue (C̲) in the C̲pG and GC̲GC sequences, respectively using S-adenosyl-L-methionine (AdoMet) as the methyl donor (Figure 2). As a result, AdoMet is transformed to S-adenosyl-L-homocysteine (AdoHcy). The sequence specificity of M.SssI is the same as that of mammalian C5 MTases. The recognition of DNA sequences and the catalytic mechanism of the C5 MTase M.HhaI have been thoroughly investigated. In fact, this enzyme has served as a paradigm of MTase activity and has been more extensively studied than any of the other C5 MTases (*19*). Taking into account that mammalian and prokaryotic C5 MTases share conserved amino acid motifs that are involved in catalysis, M.SssI and M.HhaI are the prokaryotic

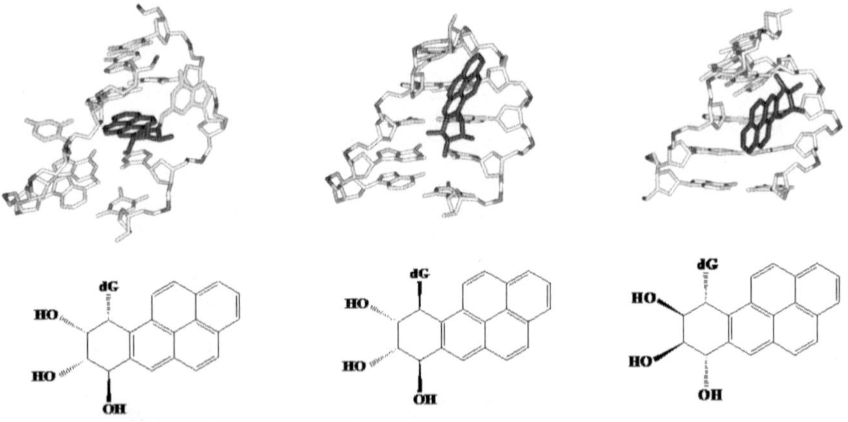

(+)-*cis*-B[*a*]P-N^2-dG (+)-*trans*-B[*a*]P-N^2-dG (-)-*trans*-B[*a*]P-N^2-dG

Figure 1. Effects of stereochemistry on the conformations of stereoisomeric B[a]P-N^2-dG adducts. The 5'-ends of the modified strands are at the top of each structure.

MTases that can be considered to be the best models of mammalian MTases. The mammalian enzyme Dnmt3a is a *de novo* MTase (*20*) that is involved in the re-methylation of the genome in early embryogenesis. The C-terminal (catalytic) domain of Dnmt3a resembles the domains of bacterial cytosine MTases and is catalytically active (*21*).

The objectives of this study were to examine the effects of adduct stereochemistry of the (+)-*trans*-B[*a*]P-N^2-dG, (+)-*cis*-B[*a*]P-N^2-dG and (-)-*trans*-B[*a*]P-N^2-dG adducts on DNA methylation by model MTases. Oligonucleotides containing these lesions positioned adjacent to, or within MTase recognition sequences, were synthesized. The effects of each of these lesions on MTase binding to DNA and the catalytic activities were explored. The effects of these B[*a*]P-N^2-dG lesions on the function of the prokaryotic C5 MTase M.EcoRII (recognition site CCA/TGG), and on catalysis by the mammalian MTase Dnmt3a are also summarized.

Design of Oligonucleotide Sequences and Positioning of Stereoisomeric B[*a*]P-N^2-dG Adducts

We synthesized site-specific 18-mer B[*a*]PDE-modified oligonucleotides with single (+)- and (-)-*trans*-B[*a*]P-N^2-dG ((+)-*trans*G* and (-)*trans*G*, respectively) and (+)-*cis*-B[*a*]P-N^2-dG adducts ((+)*cis*G*) (Figure 3). Each of these stereoisomeric lesions were introduced as single substituents into the target recognition sequences, or adjacent to these sites of the SssI, HhaI (Figure 3) (*22*)) and EcoRII (*23*) MTases, by replacement of unmodified guanine residues with (+)*trans*G*, (-)*trans*G*, or (+)*cis*G*. The methods of synthesis of these modified oligonucleotides has been described in previous publications (*22–25*). Each of these DNA duplexes contained overlapping recognition sites of M.HhaI (GCGC) and M.SssI/Dnmt3a (CpG), and were used as substrates for these MTases (Figure

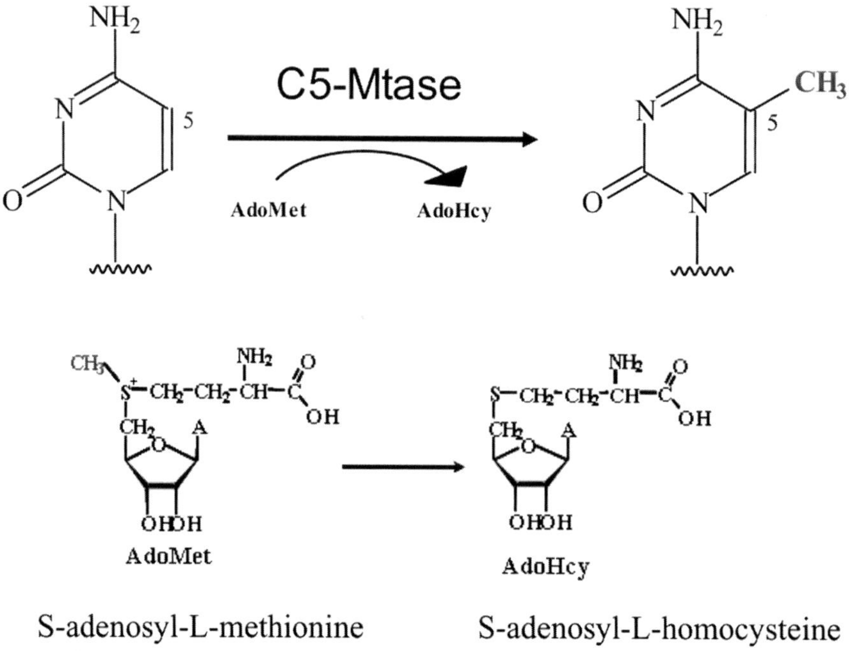

Figure 2. Methylation of the C5 position of cytosine by C5-cytosine DNA methyltransferases in the presence of the cofactor S-adenosyl-L-methionine.

3). Both unmethylated and hemi-methylated DNA duplexes were employed in our studies. The bulky substituents cause only moderate thermal destabilization of the 18-mer double-stranded DNA molecules (*22*, *23*). Thus, all duplexes containing *trans*G* or *cis*G* adducts were stable under the conditions of the methylation reactions.

Interaction of the MTases with DNA Containing Stereoisomeric Minor Groove (+)- and (-)-*trans*-B[*a*]P-*N*²-dG Adducts

Determination of Binding Affinities

The binding of M.HhaI and M.SssI to the unmodified and to the B[*a*]PDE-modified duplexes was investigated in the presence of the co-factor analog AdoHcy that facilitates the formation of the complexes (*26*). The dissociation constants, K_d, of the ternary MTase-DNA-AdoHcy complexes were determined by an electrophoretic mobility shift assay and by fluorescence polarization measurements (*22*, *23*, *27*). These minor groove lesions have virtually no effect on the binding affinity of M.SssI to the modified DNA duplexes, regardless of the position of the adduct, its stereochemistry, or the presence of ^{5Me}dC rather than the target dC residues (Figure 4). The binding affinity of the M.HhaI to B[*a*]PDE-modified hemi-methylated sequences is 1-2 orders of magnitude smaller than to the parent duplexes (Figure 4). Similar results were observed in the case of unmethylated DNA (*22*). For M.EcoRII, the K_d values are

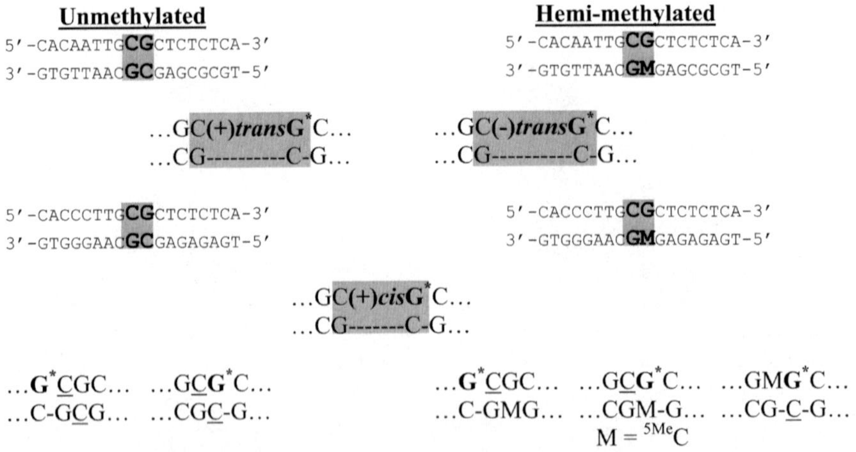

Figure 3. Benzo[a]pyrene-modified DNA duplexes as substrates of M.SssI and M.HhaI. The M.SssI recognition site is shaded in grey.

greater than those measured with unmodified duplexes by factors of 5-30 in the case of these minor groove B[*a*]P-*N*2-dG adducts (*23*).

According to the available crystal structures of complexes of M.HhaI with unmodified DNA and the co-factor, the M.HhaI consists of two domains, the large domain containing the AdoMet binding site and the catalytic center, and the small domain containing the target recognition domain (*19*, *28*). The DNA molecule is positioned in the cleft between the two domains. Before methylation, the target dC residue flips out of the DNA double helix into the M.HhaI active-site pocket (*28*). The flipped out cytosine forms contacts with the catalytic loop of the enzyme on the DNA minor-groove side. The observed decrease in binding may be explained by a perturbation of the contacts of the catalytic loop of M.Hhal with the DNA minor groove that help to stabilize the catalytic loop. According to the model of the M.Sssl·DNA·AdoHcy complex, most of the sequence-specific DNA-protein contacts are located in the major groove of the double helix (*29*). This observation suggests that the introduction of the B[*a*]P residue into the DNA minor groove does not significantly perturb the sequence specific recognition of DNA by M.SssI, and this was confirmed experimentally.

In analogy to the higher affinity of MTases for DNA containing mispaired cytosine residues (*30*), one can suppose that MTases might also exhibit a higher affinity of binding to DNA that contains B[*a*]P-*N*2-dG adducts that diminish stability of Watson-Crick base pairing. However, the B[*a*]P lesions practically do not affect the binding of M.SssI to DNA, and only moderately diminish the binding of M.Hhal and M.EcoRII to the modified DNA sequences studied. M.Hhal and M.EcoRII, and probably other prokaryotic MTases that recognize 4-8 base pair sequences in DNA appear to be more sensitive to the replacement of a nucleobase by a B[*a*]P-modified base within the recognition sequence than the CpG sequence-recognizing M.SssI.

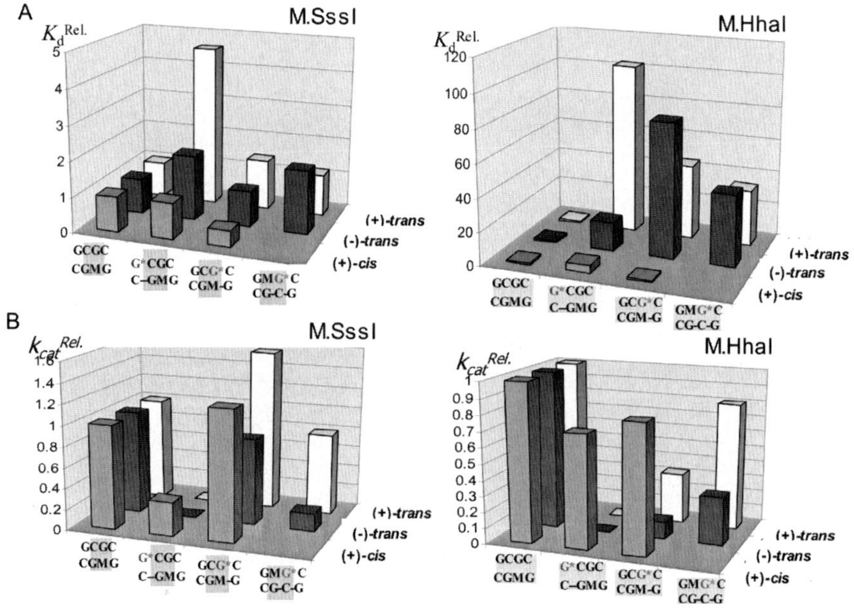

Figure 4. Bar graphs representing relative K_d (K_d^{rel}) (A) and k_{cat} (k_{cat}^{rel}) (B) values for binding and methylation of DNA containing (+)-trans-B[a]P-N^2-dG, (-)-trans-B[a]P-N^2-dG and (+)-cis-B[a]P-N^2-dG adducts by M.SssI and M.HhaI. The K_d^{rel} and k_{cat}^{rel} values for duplexes containing the B[a]P-N^2-dG adducts (G) were calculated relative to the duplex GCGC/CGMG. The M.SssI and M.HhaI recognition sites are highlighted and shaded in gray.*

Steady-State Kinetics of Methylation Catalyzed by M.SssI, M.HhaI and M.EcoRII

Methylation of the unmethylated B[*a*]P-modified duplexes by M.SssI and M.HhaI was decreased in comparison with the parent, unmodified duplex (1.3-4 fold in the case of M.HhaI, and 2-22 fold in the case M.SssI) (*22*). The methylation of hemimethylated sequences by both M.SssI and M.HhaI reveals a much stronger dependence on the position of the adduct. The most striking effect on methylation rates was observed when either of the two stereoisomeric (+)- or (-)-*trans*-B[*a*]-N^2-dG adducts flanked the target dC residue on the 5′-side (Figure 4). Methylation of these duplexes was completely abolished in the case of M.SssI, or dramatically decreased in the case of M.HhaI as compared to the unmodified hemimethylated duplex. When the B[*a*]P-N^2-dG adduct was positioned within the recognition site, 3′ adjacent to the target dC, we observed a 1.2 to 8-fold decrease of methylation by both MTases. With the (+)*trans*G* or (-)*trans*G* residues flanking the ^{5Me}dC on the 3′-side, a small decrease in methylation rates (1.3 to 6-fold) was observed in the case of both MTases, as compared to the unmodified hemimethylated duplex. In the case of M.EcoRII, the catalytic activity was either abolished or reduced 4-80-fold when the adducts were located at either G of the CCA/TGG sequence (*23*).

Effects of Adduct Stereochemistry

In all substrates, methylation catalyzed by either M.SssI, M.HhaI, or M.EcoRII, is more efficient in duplexes with the (+)-*trans*- than the (-)-*trans*-B[*a*]P-N^2-dG lesions. This effect may be related to the opposite orientations, relative to the modified guanine residues, of the bulky B[*a*]PDE residues in the minor groove (Figure 1) (*31*).

In summary, under steady-state conditions, the lesions either decrease the k_{cat} values, or block methylation completely when catalyzed by M.SssI, M.HhaI, or M.EcoRII. The greatest impact of the B[*a*]P-N^2-dG lesions on methylation by M.SssI and M.HhaI is observed when the lesion flanks the target dC residue on its 5′-side.

Analysis of Single Turnover Kinetics of Methylation by M.SssI and M.HhaI

Under single turnover conditions, as well as under steady state conditions, the k_{chem} values for the hemimethylated duplexes containing *trans*-G* 5′ adjacent to the target dC were significantly lower than for the duplexes with the adducts positioned at other sites (*22*). The level of methylation of the modified duplexes by M.SssI, even after a 2 h incubation period, was quite low.

Summarizing the results on the impact of B[*a*]P-N^2-dG adducts on the various steps in the methylation catalytic cycle, we conclude that these lesions decrease both the multiple turnover and the single turnover rates of methyl transfer. The strong inhibition of DNA methylation catalyzed by prokaryotic MTases suggests that the interaction of substrate with the catalytic center of these enzymes is strongly perturbed by the minor groove B[*a*]P residues.

To account for the most pronounced effect of B[*a*]P-N^2-dG lesions on methylation of hemimethylated duplexes catalyzed by M.HhaI, we took advantage of the known crystal structure of the ternary M.HhaI·DNA·AdoHcy complex where the target cytosine is flipped out of the duplex (*32*). We speculated that, in the case of the B[*a*]P-modified duplexes characterized by strongly reduced methylation rates, the bulky B[*a*]P-N^2-dG adducts flanking the target dC residue on its 5′-side, abolishes the important DNA-enzyme contacts that stabilize the catalytic loop of M.HhaI. These primarily involve contacts of Ile86 with the amino group of the (+)*trans*G* or (-)*trans*G* adducts (Figure 5). When (+)*trans*G* or(-)*trans*G* lesions flank the target dC on its 5′-side (Figure 5), the methyl transfer reaction catalyzed by M.SssI is blocked. In this case, the bulky polycyclic aromatic ring system probably perturbs the flipping of the target cytosine out of the helix, and/or the rearrangement of the catalytic loop of M.SssI. Since methylation is essentially inhibited even under single turnover conditions for 2 h, one may assume that flipping of the target base is impeded. Therefore, the presence of the (+)*trans*G* or (-)*trans*G* lesions on the 5′-side of the CpG dinucleotide step is especially unfavorable for cytosine methylation by prokaryotic MTases in the CpG sequence context.

Steady-State Kinetics of Methylation by Dnmt3a-CD

Taking into account that Dnmt3a-CD can methylate both DNA strands, we used hemimethylated DNA duplexes as substrates. This allowed us to examine the effects of (+)- or (-)-*trans*-B[*a*]-N^2-dG adducts on the methylation of one or the other DNA strand. Introduction of (+)- or (-)-*trans*-B[*a*]-N^2-dG adducts into DNA duplexes results in a 2-6 fold decrease of the initial rates of methylation (Figure 6). The most substantial effect was observed when a guanine residue opposite to the target cytosine was substituted by a (+)-*trans*-B[*a*]-N^2-dG adduct (duplex GCGC/C**(+)*trans*G***MG), or when the guanine adjacent to the recognition site on its 5′-side was substituted by (-)-*trans*-B[*a*]-N^2-dG (duplex **(-)*trans*G***CGC/CGMG). Positioning (+)- or (-)-*trans*-B[*a*]-N^2-dG adducts within the recognition site on the 3′-side of the target cytosine (duplexes GC**(+)*trans*G***C/CGMG) also significantly decreased the initial rates of methylation. A smaller effect on methylation rates was observed when the (+)-*trans*-B[*a*]-N^2-dG adduct replaced a guanine adjacent to the recognition site on the 5′-side (duplex **(+)*trans*G***CGC/CGMG), or when the (-)-*trans*-B[*a*]-N^2-dG adduct replaced a guanine residue opposite to the target cytosine (duplex GCGC/C**(-)*trans*G***MG). We recall that the (+)- and (-)-*trans*- B[*a*]P-N^2-dG adducts adopt a conformation that has the benzo[*a*]pyrenyl residue in the minor groove of DNA, with the (+)-*trans*-stereoisomer oriented towards the 5′-end of the modified DNA strand, and the (-)-*trans*-stereoisomer oriented toward the 3′end of the modified DNA strand (Figure 2) (*4*). The influence on DNA methylation was the most prominent when the benzo[*a*]pyrenyl residue shielded the minor groove in the CpG site region. It appears that the benzo[*a*]pyrenyl residues positioned in the minor groove perturb the interaction of the Dnmt3a-CD catalytic loop with the DNA molecule and thus cause the observed decreases in DNA methylation rates.

It is interesting to compare the influence of (+)- or (-)-*trans*-B[*a*]-N^2-dG adducts on DNA methylation catalyzed by the CpG-recognizing MTases SssI and Dnmt3a-CD. DNA methylation by M.SssI was blocked when (+)*trans*G* or (-)*trans*G* lesions flanked the target dC on its 5′-side. In the case of Dnmt3a-CD, the most profound effect on methylation was observed when the benzo[*a*]pyrenyl residue shielded the minor groove in the CpG site region. According to footprint analysis of M.SssI (*33*) and homology modeling of M.SssI in the complex with DNA and AdoHcy (*29*) the catalytic loop of this MTase interacts with the nucleoside that flanks the CpG sequence on its 5′-side. For the mammalian MTase Dnmt3a-CD, we speculate that its catalytic loop forms contacts that are critical for DNA methylation directly or close to the CpG sequence.

Interaction of MTases with (+)-*cis*-B[*a*]P-N^2-dG Adducts in DNA

We recall that this lesion has a base-displaced intercalative conformation with the modified guanine and partner C residues flipped out of the double helix (Figure 1). M.HhaI and M.SssI bind unmethylated B[*a*]P-modified DNA containing (+)*cis*G* lesions, as well as hemimethylated duplexes with (+)*cis*G* adducts (Figure 4) with affinities that are comparable to those of the corresponding

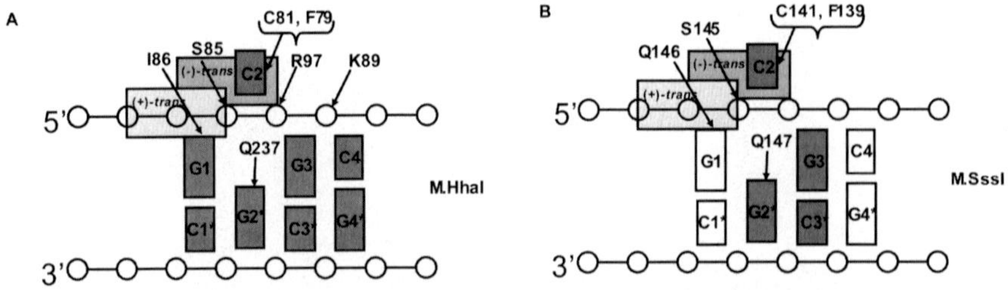

Figure 5. Schematic diagrams showing the contacts of M.HhaI (A) and M.SssI (B) with the minor groove of DNA containing (+)-trans-B[a]P-N^2-dG or (-)-trans-B[a]P-N^2-dG on the minor groove side. The nucleoside residues are represented by rectangles and the backbone phosphate groups are represented by circles. The protein residues interacting with DNA are represented by their single-letter identifiers and numbers. HhaI and SssI sites are shown in dark gray. These schemes were derived from the crystal structure of the M.HhaI·DNA·AdoHcy complex (32) and a model of the M.SssI·DNA·AdoHcy complex (29).

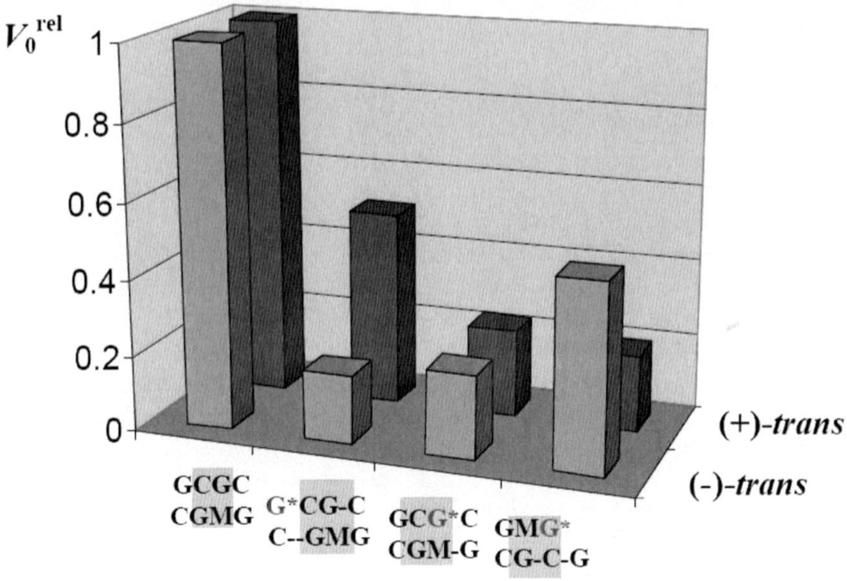

Figure 6. Bar graph representing relative V_0 (V_0^{rel}) values for methylation of DNA containing (+)-trans-B[a]P-N^2-dG or (-)-trans-B[a]P-N^2-dG adducts (G) by Dnmt3a. The V_0^{rel} values for duplexes containing G* were calculated relative to the duplex G<u>C</u>GC/CGMG. The Dnmt3a recognition site is shaded in gray.*

unmodified duplexes (*27*). M.SssI and M.HhaI transfer methyl groups with high efficiencies to all DNA duplexes containing the (+)*cis*G* adducts regardless of the position of the modified nucleoside (Figure 4). Therefore, the introduction of the intercalative, base displaced (+)-*cis*-B[*a*]P-*N*^{2}-dG adduct into double-stranded DNA has practically no effect on methylation rates catalyzed by the MTases SssI

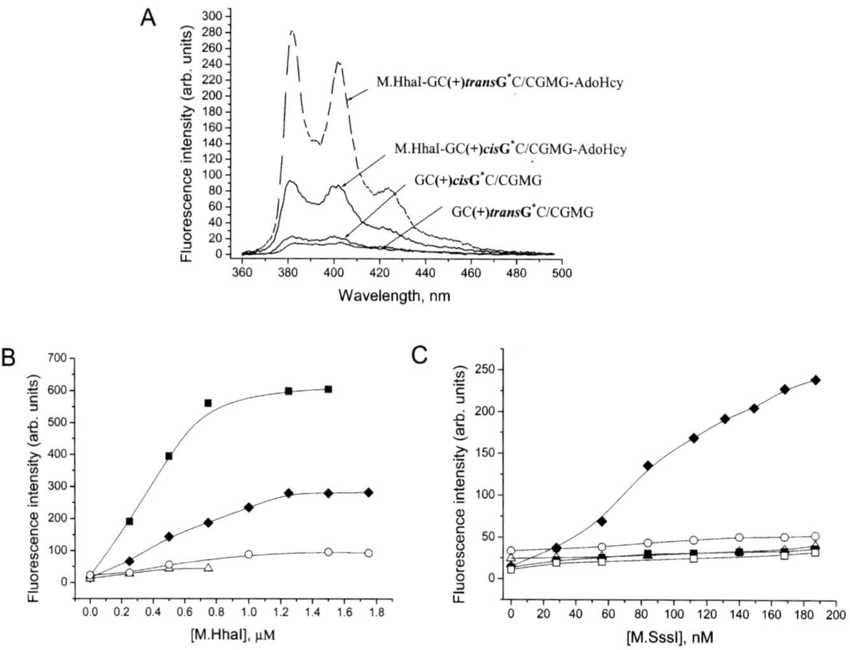

*Figure 7. Fluorescence titration of DNA containing (+)-cis-B[a]P-N^2-dG or (+)-trans-B[a]P-N^2-dG adducts with M.HhaI or M.SssI. (A) Typical fluorescence emission spectra of the M.HhaI•B[a]P-DNA•AdoHcy complexes and the free B[a]P-DNA duplexes. The fluorescence excitation wavelength was 350 nm. (B) 500 nM GC**(+)cisG***C/CGMG (○), GC**(+)transG***C/CGMG (◆), and **(+)transG***CGC/CGMG (■), or 200 nM of **(+)cisG***CGC/CGMG (Δ) were titrated with M.HhaI at 25°C and then the emission at 384 nm was measured with excitation at 350 nm. (C) 100 nM GC**(+)cisG***C/CGMG (○), **(+)cisG***CGC/CGMG (Δ), **(+)transG***CGC/CGMG (■), GC**(+)transG***C/CGMG (◆) or **(+)transG***(N)4CG/C(N)$_4$GM (□) were titrated with M.SssI at 25 °C. The excitation and emission wavelengths are the same as in (B). Reproduced with permission from reference (27). Copyright 2007 Blackwell Publishing Ltd., Oxford, UK.*

and HhaI, in contrast to the large effects observed in the case of the minor groove (+)-*trans*- and (-)-*trans*-B[*a*]P-N^2-dG adducts.

Effects of M.HhaI and M.SssI Binding on the Fluorescence of the (+)-*trans*-B[*a*]P-N^2-dG and (+)-*cis*-B[*a*]P-N^2-dG Adducts

In order to probe the changes in the immediate environment of the stereoisomeric (+)-*cis*-B[*a*]P-N^2-dG and (+)-*trans*-B[*a*]P-N^2-dG adducts that occur upon binding of the MTases, the fluorescence properties of the pyrenyl residues were examined when the DNA duplexes were titrated with various amounts of M.HhaI or M.SssI (*27*). The large increase in the fluorescence yields upon formation of the M.HhaI·GC**(+)*trans*G***C/CGMG·AdoHcy, M.HhaI·**(+)*trans*G***CGC/CGMG·AdoHcy, and M.SssI·GC**(+)*trans*G***C/CGMG·

AdoHcy complexes suggest that the (+)-*trans*-adducts are situated in a more hydrophobic environment in the protein complexes than in aqueous solution in the absence of protein (Figure 7). The enhancements in the fluorescence intensities are consistent with a change of the microenvironment of the aromatic ring system from a more polar one to a less polar one. Such changes were observed previously in the case of the fluorescence of the mononucleoside (+)-*trans*-B[*a*]P-N^2-dG adduct when water was replaced by more hydrophobic organic solvents (*34*). We observed that the change in the hydrophobicity of the local environment upon protein binding is less pronounced in the case of the (+)-*cis*-B[*a*]P-N^2-dG adduct than in the case of the (+)-*trans*-B[*a*]P-N^2-dG adduct (Figure 7). The microenvironments of the pyrenyl residues in the (+)-*cis*-B[*a*]P-N^2-dG as well as the (+)-*trans*-B[*a*]P-N^2-dG adducts in MTase·B[*a*]P-DNA complexes is less polar than it is in the case of the same duplexes dissolved in aqueous buffer solution. The observed changes of the B[*a*]P fluorescence yields are consistent with the very different B[*a*]P arrangement in the complexes of the MTases with *trans*- or *cis*-B[*a*]P-DNA.

It has been proposed that the target bases flip out of the DNA duplex and that this extrusion is an important intermediate step in the DNA methylation process catalyzed by C5 MTases (*30*). Our hypothesis is that the fluorescence intensity of the (+)-*trans* adduct depends on its position relative to that of the target dC in complexes of M.SssI with the **(+)*trans*G***CGC/CGMG, GC**(+)*trans*G***C/CGMG or **(+)*trans*G***$(N)_4$CG/C$(N)_4$GM duplexes (Figure 7C). It is well known that neighboring bases in their normal positions in DNA quench the fluorescence of the aromatic pyrenyl residue of (+)-*trans*-B[*a*]P-N^2-dG adducts positioned within oligodeoxynucleotide duplexes (*35–37*). We suggest that the observed large increase in fluorescence in the case of the (+)-*trans* adduct in the CpG duplex sequence context in the M.SssI.GC**(+)*trans*G***C/CGMG complex containing the (+)-*trans* adduct may be caused by diminished quenching by the target dC residue that is flipped out in the MTase–DNA complex. When the target dC base and the B[*a*]P-modified guanine residue are separated by four nucleotides in the **(+)*trans*G***$(N)_4$CG/C$(N)_4$GM duplex, the increase in the fluorescence intensity upon formation of the M.SssI–DNA complex is substantially smaller (Figure 7C). In the case of the **(+)*trans*G***<u>C</u>GC/CGMG duplex, when the B[*a*]P aromatic ring system is not positioned in the CpG site, but flanks the target dC on its 5′-side, the fluorescence increase is small, and the methylation of this duplex is also blocked by this adduct. These observations suggest that the normal target base conformation is not perturbed, i.e., not flipped out of the duplex, and thus it does not affect the fluorescence of the B[*a*]P residue, which remains unchanged in the complex. Overall, the enhancements in the fluorescence intensities are clearly related to changes in the local microenvironment of the B[*a*]P residues in the DNA duplexes that are consistent with a base-flipping model of the target cytosine and/or a change in the local polarity of the environment.

Conclusions and Future Studies

Our studies of prokaryotic MTases HhaI and SssI have shown that (1) the methyltransferase activities of the MTases can be diminished by B[*a*]P-N^2-dG adducts in a manner that depends on adduct stereochemistry, adduct position, and methylation status of the complementary strand in the recognition sequence ; (2) the intercalated (+)-*cis*-B[*a*]P-N^2-dG adduct, in contrast to the biologically more significant minor groove (+)-*trans*-adduct, exerts only minor effects on enzyme-DNA binding and catalytic activity; and (3) the minor groove conformation of the B[*a*]P residues of this adduct leads to a perturbation of the normal contacts of the methyltransferase catalytic loop with the B[*a*]PDE-modified DNA. It appears that the same mechanism accounts for the decrease of DNA methylation by the mammalian MTase Dnmt3a-CD in the case of bulky B[*a*]P residues positioned in the minor groove.

Our results demonstrate that the stereoisomeric B[*a*]P-N^2-dG adduct can, in addition to mutations, lead to a decreased efficiency of methylation of the B[*a*]PDE-damaged DNA. There are a number of important unresolved questions that remain open. These include the demonstration using site-specifically modified DNA adducts that these phenomena can also occur in vivo, and whether the observed impact of the B[*a*]P-DNA adducts on MTase activity is unique or typical of other PAH-DNA adducts as well. Further studies of the interactions of mammalian Mtases with site-specific PAH adducts in DNA are currently being undertaken to provide new information on the impact of these lesions on DNA methylation in eukaryotic cells.

Acknowledgments

This research was supported by the Civilian Research Defense Foundation/ Russian Foundation for Basic Research, Award No. RUB1-2919-MO-07/08-04-91109 to E.S.G., Russian Foundation for Basic Research Award No. 07-04-00583 to E.S.G., and by a grant from the National Cancer Institute, National Institutes of Health, Grant CA099194 to N.E.G. The content is solely the responsibility of the authors and does not necessarily represent the official views of the National Cancer Institute or the National Institutes of Health.

References

1. Luch, A. *Nat. Rev. Cancer* **2005**, *5*, 113–125.
2. Conney, A. H. *Cancer Res.* **1982**, 4875–4917.
3. Cheng, S. C.; Hilton, B. D.; Roman, J. M.; Dipple, A. *Chem. Res. Toxicol.* **1989**, *2*, 334–340.
4. Geacintov, N. E.; Cosman, M.; Hingerty, B. E.; Amin, S.; Broyde, S.; Patel, D. J. *Chem. Res. Toxicol.* **1997**, *10*, 111–146.
5. Cosman, M.; de los Santos, C.; Fiala, R.; Hingerty, B. E.; Singh, S. B.; Ibanez, V.; Margulis, L. A.; Live, D.; Geacintov, N. E.; Broyde, S.; Patel, D. J. *Proc. Natl. Acad. Sci. U.S.A.* **1992**, *89*, 1914–1918.

6. de los Santos, C.; Cosman, M.; Hingerty, B. E.; Ibanez, V.; Margulis, L. A.; Geacintov, N. E.; Broyde, S.; Patel, D. J. *Biochemistry* **1992**, *31*, 5245–5252.
7. Cosman, M.; de los Santos, C.; Fiala, R.; Hingerty, B. E.; Ibanez, V.; Luna, E.; Harvey, R.; Geacintov, N. E.; Broyde, S.; Patel, D. J. *Biochemistry* **1993**, *32*, 4145–4155.
8. Kaufmann, W. K. *Cell Cycle* **2007**, *6*, 1460–1467.
9. Fernandes, A.; Liu, T.; Amin, S.; Geacintov, N. E.; Grollman, A. P.; Moriya, M. *Biochemistry* **1998**, *37*, 10164–10172.
10. Avkin, S.; Goldsmith, M.; Velasco-Miguel, S.; Geacintov, N.; Friedberg, E. C.; Livneh, Z. *J. Biol. Chem.* **2004**, *279*, 53298–53305.
11. Seo, K. Y.; Jelinsky, S. A.; Loechler, E. L. *Mutat. Res.* **2000**, *463*, 215–246.
12. Denissenko, M. F.; Chen, J. X.; Tang, M. S.; Pfeifer, G. P. *Proc. Natl. Acad. Sci. U.S.A.* **1997**, *94*, 3893–3898.
13. Chen, J. X.; Zheng, Y.; West, M.; Tang, M. S. *Cancer Res.* **1998**, *58*, 2070–2075.
14. Weisenberger, D. J.; Romano, L. J. *J. Biol. Chem.* **1999**, *274*, 23948–23955.
15. Hess, M. T.; Gunz, D.; Luneva, N.; Geacintov, N. E.; Naegeli, H. *Mol. Cell. Biol.* **1997**, *17*, 7069–7076.
16. Mocquet, V.; Kropachev, K.; Kolbanovskiy, M.; Kolbanovskiy, A.; Tapias, A.; Cai, Y.; Broyde, S.; Geacintov, N. E.; Egly, J. M. *EMBO J.* **2007**, *26*, 2923–2932.
17. Wilson, V. L.; Smith, R. A.; Longoria, J.; Liotta, M. A.; Harper, C. M.; Harris, C. C. *Proc. Natl. Acad. Sci. U.S.A.* **1987**, *84*, 3298–3301.
18. Wojciechowski, M. F.; Meehan, T. *J. Biol. Chem.* **1984**, *259*, 9711–9716.
19. Sankpal, U. T.; Rao, D. N. *Crit. Rev. Biochem. Mol. Biol.* **2002**, *37*, 167–197.
20. Jeltsch, A. *Curr. Top. Microbiol. Immunol.* **2006**, *301*, 203–225.
21. Gowher, H.; Jeltsch, A. *J. Biol. Chem.* **2002**, *277*, 20409–20414.
22. Subach, O. M.; Baskunov, V. B.; Darii, M. V.; Maltseva, D. V.; Alexandrov, D. A.; Kirsanova, O. V.; Kolbanovskiy, A.; Kolbanovskiy, M.; Johnson, F.; Bonala, R.; Geacintov, N. E.; Gromova, E. S. *Biochemistry* **2006**, *45*, 6142–6159.
23. Baskunov, V. B.; Subach, F. V.; Kolbanovskiy, A.; Kolbanovskiy, M.; Eremin, S. A.; Johnson, F.; Bonala, R.; Geacintov, N. E.; Gromova, E. S. *Biochemistry* **2005**, *44*, 1054–1066.
24. Johnson, F.; Bonala, R.; Tawde, D.; Torres, M. C.; Iden, C. R. *Chem. Res. Toxicol.* **2002**, *15*, 1489–1494.
25. Geacintov, N. E.; Cosman, M.; Mao, B.; Alfano, A.; Ibanez, V.; Harvey, R. G. *Carcinogenesis* **1991**, *12*, 2099–2108.
26. Subach, O. M.; Khoroshaev, A. V.; Gerasimov, D. N.; Baskunov, V. B.; Shchyolkina, A. K.; Gromova, E. S. *Eur. J. Biochem.* **2004**, *271*, 2391–2399.
27. Subach, O. M.; Maltseva, D. V.; Shastry, A.; Kolbanovskiy, A.; Klimasauskas, S.; Geacintov, N. E.; Gromova, E. S. *FEBS J.* **2007**, *274*, 2121–2134.
28. Klimasauskas, S.; Kumar, S.; Roberts, R. J.; Cheng, X. *Cell* **1994**, *76*, 357–369.

29. Koudan, E. V.; Bujnicki, J. M.; Gromova, E. S. *J. Biomol. Struct. Dyn.* **2004**, *22*, 339–346.
30. Roberts, R. J.; Cheng, X. *Annu. Rev. Biochem.* **1998**, *67*, 181–198.
31. Yan, S.; Wu, M.; Patel, D. J.; Geacintov, N. E.; Broyde, S. *Biophys. J.* **2003**, *84*, 2137–2148.
32. O'Gara, M.; Klimasauskas, S.; Roberts, R. J.; Cheng, X. *J. Mol. Biol.* **1996**, *261*, 634–645.
33. Renbaum, P.; Razin, A. *J. Mol. Biol.* **1995**, *248*, 19–26.
34. Geacintov, N. E.; Mao, B.; France, L. L.; Zhao, R.; Chen, J.; Liu, T.-M.; Ya, N.-Q.; Margulis, L. A.; Sutherland, J. C. *SPIE Time-Resolved Laser Spectroscopy Biochemistry III* **1992**, *1640*, 774–783.
35. Geacintov, N. E.; Cosman, M.; Mao, B.; Alfano, A.; Ibanez, V.; Harvey, R. G. *Carcinogenesis* **1991**, *12*, 2099–2108.
36. Mao, B.; Xu, J.; Li, B.; Margulis, L. A.; Smirnov, S.; Ya, N. Q.; Courtney, S. H.; Geacintov, N. E. *Carcinogenesis* **1995**, *16*, 357–365.
37. Ponten, I.; Kim, S. K.; Graslund, A.; Norden, B.; Jernstrom, B. *Carcinogenesis* **1994**, *15*, 2207–2213.

Indexes

Author Index

Subject Index

E

F

G

H

M

N

O

P

R

S

T

U

X